SCIENTIFIC AMERICAN EXPLORES BIG IDEAS

Smart Cities of the Future

The Editors of *Scientific American*

SCIENTIFIC AMERICAN EDUCATIONAL PUBLISHING

New York

Published in 2024 by Scientific American Educational Publishing
in association with **The Rosen Publishing Group**
2544 Clinton Street, Buffalo NY 14224

Contains material from Scientific American®, a division of Springer Nature America, Inc., reprinted by permission, as well as original material from The Rosen Publishing Group®.

First Edition

Scientific American
Lisa Pallatroni: Project Editor

Rosen Publishing
Michael Hessel-Mial: Compiling Editor
Michael Moy: Senior Graphic Designer

Cataloging-in-Publication Data
Names: Scientific American, Inc.
Title: Smart cities of the future / edited by the Scientific American Editors.
Description: First Edition. | New York : Scientific American Educational Publishing, 2024. | Series: Scientific American explores big ideas | Includes bibliographic references and index.
Identifiers: ISBN 9781725349605 (pbk.) | ISBN 9781725349612 (library bound) | ISBN 9781725349629 (ebook)
Subjects: LCSH: Smart cities–Juvenile literature. | Cities and towns–Juvenile literature.
Classification: LCC TD159.4 S637 2024 | DDC 307.760285–dc23

Manufactured in the United States of America
Websites listed were live at the time of publication.

Cover: metamorworks/Shutterstock.com

CPSIA Compliance Information: Batch # SACS24.
For Further Information contact Rosen Publishing at 1-800-237-9932.

CONTENTS

INTRODUCTION

You might have seen a number of videos depicting some aspect of city life: plans for gleaming future cities in the Middle East, former factories or rail lines converted into green spaces, enormous skyscrapers in Asia, flooded main streets after natural disasters, urban poverty from the rising cost of living. Cities remain a significant topic of conversation for their potential and for their shortcomings.

The fact is, city life has now become the norm for the majority of the world's population, and that number is expected to become a supermajority by the end of the century. While U.S. and European cities have grown slowly or shrank in past decades, Asian cities have expanded, creating comprehensive transit systems and tall, dense spaces to work and play. Other cities from poorer nations highlight another model of growth: bustling urban cores surrounded by underserved slums, whose dwellers must rely on their own ingenuity to create shelter, community, and opportunity.

Many wonder what cities might look like in the future, especially given the hard environmental and social challenges we face. It turns out, scientists have spent time on these questions. Section 1, "Trends in Urbanization," gives background on cities past and present. Section 2, "The Science of Cities," explores cities as systems whose complexity drives innovation and growth. Section 3, "Urban Inequality," discusses the social inequality that comes with urbanization, from the slums of Rio de Janeiro to the gentrification struggles in Brooklyn, New York. Section 4, "Imagining the Future City," explores "smart" cities, car-free cities, and other ambitious ideas. Section 5, "Designing the Green City," takes on environmental challenges all cities face, while showing. how cities have unique opportunities for sustainability.

Whether you live in a small city like Billings, Montana, or a megacity like Jakarta, Indonesia, your urban experience has features that a medieval village dweller might find incomprehensible. As our world grows more urbanized, we too might find ourselves surprised by what comes next.

Section 1: Trends in Urbanization

Myths of the City

By Roger Doyle

From the beginning of the Christian era to about 1850, the urban population of the world never exceeded 7 percent. The Industrial Revolution quickly changed that–today 75 percent of people in the U.S. and other developed countries live in cities, according to the United Nations.

Urbanization in the developing countries (such as China and India) and the least developed countries (such as Ethiopia and Bangladesh) has long lagged behind that of the West and Japan. Early in the 20th century probably no more than 5 percent of the population in European colonies clustered in cities. But since then, the proportion in these ex-colonies and the never-colonized nations has increased twice as fast as that of the West. New York City, for instance, took about 150 years to add eight million, but Mexico City added 16 million in just 50 years. Such extraordinary growth has exacerbated the usual problems of development, making it difficult to supply drinking water, sewage disposal, police protection and other amenities. The comparatively slow pace of European urbanization was partly the result of 45 million or so people emigrating to the New World rather than moving to cities.

An enduring myth is that the country is healthier than the city. Historically that was true, but no longer. Urban sanitary measures and ready access to good medical care have made the cities healthier. Another myth is that city people are isolated. But a study by sociologists Katherine J. Curtis White and Avery M. Guest of the University of Washington found no disparity in the strength of social ties.

Most of the ills ascribed to urbanization stem from population growth, industrialization and prosperity, which would continue to cause economic dislocation even if urbanization were reversed. Many problems reflect rapid growth rather than numbers of people per se, and in some developing countries, such as China, the wretched condition of factory workers can be blamed on lack of social justice, not the growth of the city.

Manufacturing productivity increases with metropolitan size in part because of publicly funded streets, highways, gas, water, sewerage, power grids, and police and fire stations. Economy of scale makes urban services relatively cheap. Indeed, megacities contribute disproportionately to national economies, as in the case of São Paulo, which has 10 percent of Brazil's population but contributes at least 25 percent of the net national product.

The U.N. forecasts a continuing increase in world urbanization over the next quarter of a century to 61 percent. If the developing countries match the record of the West and Japan, the world will eventually reach a level of over 80 percent. New data from the Earth Institute at Columbia University show that there is more than enough room for expansion: urban areas now cover only 3 percent of the planet's land surface.

Lost Garden Cities: Pre-Columbian Life in the Amazon

By Michael J. Heckenberger

When Brazil established the Xingu Indigenous Park in 1961, the reserve was far from modern civilization, nestled deep in the southern reaches of the vast Amazon forest. When I first went to live with the Kuikuro, one of the reserve's principal indigenous groups, in 1992, the park's boundaries were still largely hidden in thick forest, little more than lines on a map. Today the park is surrounded by a patchwork of farmland, its borders often marked by a wall of trees. For many outsiders, this towering green threshold is a portal, like the massive gates of Jurassic Park, between the present—the dynamic modern world of soy fields, irrigation systems and 18-wheelers—and the past, a timeless world of primordial nature and society.

Long before taking center stage in the world's environmental crisis as the giant green jewel of global ecology, the Amazon held a special place in the Western imagination. Mere mention of its name conjures images of dripping, vegetation-choked jungles; cryptic, colorful and often dangerous wildlife; endlessly convoluted river networks; and Stone Age tribes. To Westerners, Amazonian peoples are quintessential simple societies, small groups that merely make do with what nature provides. They have complex knowledge about the natural world but lack the hallmarks of civilization: centralized government, urban settlements and economic production beyond subsistence. In 1690 John Locke famously proclaimed, "In the beginning all the World was America." More than three centuries later the Amazon still grips the popular imagination as nature at its purest, home to native peoples who, in the words of *Rolling Stone* editor Sean Woods in October 2007, preserve "a way of life unchanged since the dawn of time."

Looks can be deceiving. Hidden under the forest canopy are the remnants of a complex pre-Columbian society. Working with the

Kuikuro, I have excavated a network of ancestral towns, villages and roads that once supported a population perhaps 20 times its present size. Huge swaths of forest have grown over the ancient settlements, gardens, fields and orchards, which fell into disuse when epidemics brought by European explorers and colonists decimated the native peoples. The region's rich biodiversity reflects past human intervention. By developing a mix of land uses, soil-enrichment techniques and long crop rotation cycles, the ancestors of the Kuikuro thrived in the Amazon despite its infertile natural soils. Their accomplishments could inform efforts to reconcile the environmental and development goals of this region and other parts of the Amazon.

"Nature Folk"

The most famous person to go looking for lost civilizations in the southern Amazon was Percy Harrison Fawcett. The British adventurer scoured what he called the "uncharted jungles" for an ancient city, Atlantis in the Amazon, replete with stone pyramids, cobbled streets and alphabetic writing. His tales inspired Conan Doyle's *The Lost World* and perhaps the Indiana Jones movies. David Grann's gripping recent book, *The Lost City of Z*, retraced Fawcett's path before his disappearance in the Xingu in 1925.

Actually, five German expeditions had already visited the Xinguano people and lands. In 1894 Karl von den Steinen's book *Unter den Naturvölkern Zentral Brasiliens*, which described the earliest expeditions, became an instant classic in the fledgling discipline of anthropology. The book set the tone for 20th-century studies of Amazonian peoples as small, isolated groups living in a delicate balance with the tropical forest: "nature folk." Later anthropologists often viewed the forest environment as uniformly inimical to agriculture; the soil's poor fertility seemed to preclude large settlements or dense regional populations. By this reasoning, the Amazon of the past must have looked much like the Amazon in recent times.

But this view began to erode in the 1970s as scholars revisited early European accounts of the region, which talked not of small tribes but of dense populations. As Charles Mann's best-selling book *1491* has eloquently described, the Americas were heavily populated on the eve of the European landings, and the Amazon was no exception. Gaspar de Carvajal, the missionary who chronicled the first Spanish expedition down the river, noted fortified towns, broad, well-kept roads and large numbers of people. Carvajal wrote on June 25, 1542:

> We went among some islands which we thought uninhabited, but after we got to be in among them, so numerous were the settlements which came into sight ... that we grieved ... and, when they saw us, there came out to meet us on the river over two hundred pirogues [canoes], that each one carries twenty or thirty Indians and some forty ... they were colorfully decorated with various emblems, and they had with them many trumpets and drums ... and on land a marvelous thing to see were the squadron formations that were in the villages, all playing instruments and dancing about, manifesting great joy upon seeing that we were passing beyond their villages.

Archaeological research in several areas along the Amazon River, such as Marajó Island at the mouth of the river and sites near the modern cities of Santarém and Manaus, has confirmed these accounts. These societies interacted in far-flung systems of trade. Less is known about the southern peripheries of the Amazon, but recent work in Llanos de Mojos in lowland Bolivia and in the Brazilian state of Acre suggests that they, too, supported complex societies. In 1720 Brazilian frontiersman António Pires de Campos described a densely settled landscape in the headwaters of the Tapajós River, just west of the Xingu:

> These people exist in such vast quantity, that it is not possible to count their settlements or villages, [and] many times in one day's march one passes ten or twelve villages, and in each one of them there are ten to thirty houses, and in these houses there are some that are thirty to forty paces across ... even their roads they make very straight and wide, and they keep them so clean that one finds not even a fallen leaf....

An Ancient Walled Town

When I ventured to Brazil in the early 1990s to study the deep history of the Xingu, lost cities were the furthest things from my mind. I had read Steinen but had barely heard of Fawcett. Although much of the vast Amazon basin was archaeological terra incognito, it was unlikely that ethnographers, much less local Xinguanos, had missed a large monolithic center towering over the tropical forests.

Nevertheless, signs of something more elaborate than present-day settlements were all around. Robert Carneiro of the American Museum of Natural History in New York City, who lived with the Kuikuro in the 1950s, had suggested that their settled way of life and productive agricultural and fishing economy could support communities 1,000 to 2,000 strong—several times the contemporary population of a few hundred. He also cited evidence that indeed it once had: a prehistoric site (designated "X11" in our archaeological survey) that was surrounded by extensive ditches. The Villas Boas brothers—Brazilian *indigenistas* who were nominated for a Nobel Peace Prize for their part in creating the Xingu park—had reported such earthworks near many villages.

In January 1993, soon after I arrived in the Kuikuro village, the principal hereditary chief, Afukaka, took me to one of the ditches at a site (X6) they call Nokugu, named for the jaguar spirit being thought to live there. We passed local men who were raising a huge fish weir across the Angahuku River, which was already swelling from the seasonal rains. The ditch, which runs over two kilometers, was two to three meters deep and more than 10 meters wide. Even though I had expected to find an archaeological landscape different from today's, the scale of these ancient communities and their constructions surprised me. Kuikuro research assistants and I spent the following months mapping it and other earthworks at the 45-hectare site.

Since that time, our team has studied numerous other sites in the area, hacking more than 30 kilometers of line-of-sight transects through the forest to map, examine and excavate the sites. Many

Kuikuro helped in one way or another, and some became well versed in archaeology.

At the end of 1993, Afukaka and I went back to Nokugu so I could tell him what I had learned. We followed the contour of the site's outer ditch and stopped at an earthen bridge, where a major road we had uncovered passed over it. I pointed down the arrow-straight ancient dirt road, which was 10 to 20 meters wide and led to another ancient site, Heulugihïtï (X13), about five kilometers away. We crossed the bridge and entered Nokugu.

The road, defined by low earthen curbs, widened to 40 meters—the size of a modern four-lane highway. After a couple of hundred meters, we passed over the inner ditch and stopped to look at our recently finished excavation trench, where we had found a funnel-shaped footing for a tree-trunk palisade. Afukaka told me a story of palisaded villages and raids in his people's distant past.

As we moved farther into the ancient town, we passed through patches of forest, scrub and open areas that now cover the site—the footprints of diverse past activities. We emerged into a grassy glade of towering palms marking the former plaza. I slowly spun and pointed along the perfectly circular edge of the plaza, marked by a meter-high mound. The tall palms, I told him, had colonized the plaza centuries ago from compost gardens in domestic areas.

Leaving the plaza to explore the surrounding neighborhoods, we came across large refuse middens that closely resembled the one behind Afukaka's own house. They were filled with broken pots that he noted were exactly like those his wives used to process and cook manioc, down to minute details. On a later visit, when we were excavating a pre-Columbian house, the chief bent down in the central kitchen area, popped out a big hunk of pottery, and corroborated my sense that the daily life of the ancient society was much like today's. "You're right!" Afukaka exclaimed. "Look here, a pot support"—an *undagi*, as the Kuikuro call it, used to cook manioc.

These connections are what make the Xinguano sites so fascinating. They are among the few pre-Columbian settlements in the Amazon where archaeological evidence can be linked directly to present-day customs. Elsewhere, the indigenous culture was completely wiped out, or the archaeological record is spotty. The ancient walled town I showed Afukaka was much like his current village, with its central plaza and radial roads, only it was 10 times larger.

From House to Polity

"Palatial" is not the word that usually comes to mind to describe a pole-and-thatch house. Most Westerners think "hut." But the house that the Kuikuro were building for the chief when I arrived in 1993 was massive: well over 1,000 square meters. It is hard to imagine that a house built like a giant, overturned basket without stone, mortar or nails could get any bigger. Even the average Xinguano house, at 250 square meters, is as big as the average American home.

What makes the chief's house stand out is not just size but also its position, located on the southern point of the central circular plaza. As one enters the village along the formal entry road, high-ranking families live to the right (south) and left (north). The arrangement reproduces, on a larger scale, the layout of an individual house, whose highest-ranking occupant hangs his hammock to the right, along the long axis of the house. The entry road runs approximately east-west; in the chief's house, his hammock is oriented in the same direction. When a chief dies, he is also laid to rest in a hammock with his head to the west.

This basic corporeal calculus applies on all scales, from houses to the entire Upper Xingu basin. Ancient towns are distributed across the region and interconnected by a lattice of precisely aligned roads. When I first arrived in the area, it took weeks to map the ditches, plazas and roads using standard archaeological techniques. Beginning in 2002, we began using precise GPS, enabling us to map major earthworks in a matter of days. What we have found is an impressive degree of regional integration. The landscape planning seems almost overdetermined, with a specific place for everything. Yet it was based on the same basic principles of the current village.

Main roads run east-west, secondary roads radiate out to the north and south, and smaller roads proliferate in other directions.

We mapped two hierarchical clusters of towns and villages in our study area. Each consists of a major ceremonial center and several large satellite towns in precise orientations relative to the center. These towns likely held 1,000 or more inhabitants. Smaller villages are located farther from the center. The northern cluster is centered on X13, which is not a town so much as a ritual center, rather like a fairground. Two large walled settlements lie equidistant to the north and south of X13, and two medium-size walled towns lie equidistant to the northeast and southwest. The southern cluster is slightly different. It is centered on X11, which is both a ritual center and a town, around which are medium- and small-size plaza settlements.

In land area, each cluster was more than 250 square kilometers, of which about a fifth was the built-up core area, making it roughly equivalent in size to a small modern city. Today most of the ancient landscape is overgrown, but forests in the core areas have distinctive concentrations of certain plants, animals, soils and archaeological artifacts, such as prolific ceramics. Land use was more intensive in the past, but the remains suggest that many practices were similar to those of the Kuikuro: manioc plots, small orchards of pequi fruit trees and fields of *sapé* grass, the preferred material for house thatch. The countryside was a patchy landscape interspersed with secondary forests that invaded fallow agricultural areas. Wetlands, which today are choked with Buriti palm, the most important industrial crop, preserve diverse evidence of fish farming, such as artificial ponds, raised causeways and weir footings. Outside the core areas was a more lightly populated green belt and even deeper forest wilderness between clusters. This forest, too, had its uses for animals, medicinal plants and certain trees, and it was considered the home of diverse forest spirits.

The areas in and around residential sites are marked by dark earth, which the Kuikuro call *egepe*, a highly fertile soil that has been enriched by household refuse and specialized soil-management activities, such as controlled burning of vegetation cover. People have altered soils the world over, making them darker, more loamy and richer in certain chemicals. In the Amazon these changes are particularly important for agriculture in many areas because the natural soil is so poor. In the Xingu, the dark earth is less prevalent than some areas, because local populations depend mostly on manioc and orchards, which do not require high-fertility soils.

Identification of large walled settlements over an area about the size of Vermont suggests that at least 15 clusters were spread across the Upper Xingu. But most of the region is unstudied, so the true number could have been much higher. Radiocarbon dating of our excavated sites suggests that ancestors of the Xinguanos settled the area, most likely from the west, and began to mold the forests and wetlands to their design about 1,500 years ago or before. In the centuries before Europeans first discovered the Americas, the communities were re-formed into hierarchical clusters. Records date back only to 1884, so the settlement patterns are our only way of estimating the pre-Columbian population; the scale of the clusters suggests a regional population many times larger than today, perhaps numbering 30,000 to 50,000.

Garden Cities of the Amazon

A century ago Ebenezer Howard's *Garden Cities of To-morrow* proposed a model for low-density, sustainable urban growth. A forerunner of today's green movement, Howard envisioned networked towns as an alternative to an industrial world filling with high-rise cities. Ten towns with tens of thousands of people, he suggested, could have the same functional and administrative capacity of a single megacity.

The ancient Xinguanos built such a system, a flat, green style of urbanism or proto-urbanism: an inchoate garden city. Perhaps Percy Fawcett was in the right place but looking for the wrong thing: stone cities. What the small-scale centers lacked in size and elaborate structures, they made up for in numbers and integration. Had Howard known of them, he might have devoted a passage to the "Garden Cities of Yesterday." The common conception of the city as a dense grid of masonry buildings dates to early desert oasis civilizations such as Mesopotamia but was uncharacteristic of many other environments. Not only the Amazon's tropical forests but also temperate forest landscapes throughout much of medieval Europe were dotted with towns and villages of similar size to those in the Xingu.

These insights are especially important today as the southern Amazon is redeveloped, this time by Western civilization. The transitional forest of the southern Amazon is being quickly converted into farmland and pasture. At the present rate, it will be reduced to 20 percent of its original size over the next decade. Much of what is left will be restricted to reserves, such as the Xingu, where indigenous people are the stewards of the remaining biodiversity. In these areas, saving tropical forests and protecting indigenous cultural heritage are, in many respects, one and the same thing.

About the Author

Michael J. Heckenberger has done archaeology in the Xingu region and elsewhere in the Brazilian Amazon since 1992, most recently as a professor at the University of Florida. His work focuses on the social and political organization and historical ecology of late prehistoric complex societies in the region, as well as changes in indigenous societies following European colonization not just in Brazil but also in northeastern South America, the Caribbean and northeastern North America. At home, he relaxes by cultivating bonsai.

Gigalopolises: Urban Land Area May Triple by 2030

By David Biello

More than half of the world's expected nine billion people will live in giant urban expanses by 2030 as cities and their hinterlands occupy an additional 1.2 million square kilometers, thereby tripling in size. That's an additional 1.35 billion people living in cities, suggesting that urban areas that currently occupy roughly 3 percent of the planet's surface will continue to expand. By comparison, urban areas increased by just 58,000 square kilometers between 1970 and 2000.

In new work published in *Proceedings of the National Academy of Sciences*, urban environment researcher Karen Seto of the Yale School of Forestry and Environmental Studies and her colleagues first divided the global land area into discrete parcels and, using predicted gross domestic product growth, population growth and urban land area cover in 2000, they projected which parcels had a high or low probability of succumbing to citification over the next few decades. Using that model, 1.2 million square kilometers of land have probabilities higher than 75 percent of becoming citified and nearly six million square kilometers have some probability of going urban.

"More than half of the urban land cover on the planet by 2030 has yet to be built," Seto explains. "The expansion of urban areas will have a direct impact on biodiversity hot spots."

Fifty-five percent of that expansion would come from massive urbanization in India and China—a trend that has been growing in recent decades. For example, a megalopolis similar to the urban corridor between Boston and Washington, D.C., in the U.S. is likely to form between Hangzhou and Shenyang in China. But the fastest urbanization is predicted to occur in newly developing regions in Africa, such as the coast of west Africa along the Gulf of Guinea and the shores of Lake Victoria farther south, encompassing Burundi, Kenya, Rwanda and Uganda, among other regions.

This may be bad news, in some cases, for the rich array of plants, animals and microscopic life that also inhabit Earth. The Eastern Afromontane, Guinean forests of west Africa and Western Ghats of India, along with Sri Lanka, are all home to such biodiversity—as well as projected to undergo rapid urban expansion that will encroach on the territory of already endangered amphibians, birds and mammals. Seto and her colleagues project the worst impacts of urban growth to occur in Central and South America.

In addition, such land use change is likely to result in even more of the greenhouse gas emissions that drive climate change, according to the new analysis. An estimated 1.38 billion metric tons of carbon could be released as forest transforms to roads, buildings and homes. As it stands, the world's cities bear responsibility for at least 70 percent of global carbon dioxide emissions. "We need to be more deliberate as a society about how we want urban places to become what they serve for humanity," Seto argues. "Too often, urban expansion is haphazard."

This is not just bad news for animals or the atmosphere, of course. Haphazard urbanization can also ill serve human inhabitants cut off from supplies of clean water or food. And the impact of urbanites is not confined to city limits; a typical Australian from Melbourne or Sydney requires greenhouse gas emissions, water diversion and land use from the entire continent. "Cities have always relied on their hinterlands and other distal places for resources from food and fuel to waste assimilation," the researchers wrote.

Blockading urban sprawl may not be the answer either. Allowing for the same population growth without urbanization might have even more severe environmental consequences, as subsistence farming expands, enmeshing residents in chronic and crippling poverty. If urbanization fails in developing regions, whether in Africa or the Indian subcontinent, the world could see continued migrations to London, Los Angeles and other developed world megacities, suggests physicist Luis Bettencourt of the Sante Fe

Institute, who studies urbanization issues. Indeed, Seto and her colleague's modeling suggests that cities in North America will nearly double their real estate by 2030.

The next 20 years or so represent a window of opportunity to learn how to grow. Seto, for her part, suggests more research is needed on "what types of urbanization minimize negative environmental and social impacts and how do they emerge?" Her group plans to continue its work by projecting global density of population—a key metric for minimizing environmental impacts. In addition, the world will need to avoid the kind of infrastructure lock-in—from coal-fired power plants that produce cheap electricity but also pollute to living and working areas separated by vast distances requiring energy-intensive commutes—that has characterized the first century of rapid urbanization. Some estimates suggest that as much as $30 trillion in infrastructure investments will be made to support such urbanization by 2030.

Of course, as Bettencourt observes: "Cities were never formed and have never grown with the objective of saving energy or protecting the environment." Yet, cities often serve that purpose by default. One of the defining features of slums is densely packed inhabitants, and more developed city areas often boast high land values that then encourage tall buildings also densely packed with people and businesses, which further encourage walking or mass transit. "This is how larger cities manage to do more with less," he adds, suggesting that such opportunities should be "consciously and systematically seized as development happens."

About the Author

David Biello is a contributing editor at Scientific American.

Decoding Mexico's City of Gods

By Erik Vance

Sometime in the 14th century, the first Mexica found their way into the valley of Teotihuacán. The Mexica (often incorrectly called Aztecs in modern times) were new to the region. An aggressive, ambitious people from the north, they were fast becoming the dominant force in highland Mexico, conquering territory and setting up the powerful city of Tenochtitlán, which would soon rule a massive empire from what is now called Mexico City. Imagine that first search party—bold, feeling invincible as a nascent superpower—coming into a lush green expanse surrounded by rolling hills. The warriors are following tales told by the local Toltec tribes of a place in the mountains, just 25 miles from their new home, where the gods once lived. Then, turning a bend, their bravado gives way to awe as the home of the gods looms into view. Ruins of pyramids as high as 20 stories—so big they are initially mistaken for hills—line a huge road. Everywhere the explorers look lie crumbling temples, marketplaces and relics of a long-dead civilization with no name, no writing, no history. Just a vast city, once glorious beyond imagination, now abandoned.

The Mexica eventually patterned Tenochtitlán after this ghost of a city and turned the ruins into a sort of summer retreat for the elite. They named that ancient road the Avenue of the Dead and the two biggest pyramids the Pyramids of the Sun and Moon. The old city itself they called Teotihuacán—the place where gods are born.

Some 200 years later, in 1521, Spanish conquistadors toppled the Mexica's empire. And for hundreds of years Teotihuacán rested in gentle decay. When archaeologists first began excavating the site in earnest in the early 1900s, they had no better idea of who built the place than the Mexica had. Many thought it was a minor settlement erected by scattered tribes that were later subsumed by invaders. Today scholars know that Teotihuacán is actually much older and much more important than any of those early investigators could have guessed. It was the heart of an extensive empire that predates all the highland civilizations in the region and that reached as many as 700 miles away. A city that rivaled, and perhaps even dominated, the mighty Maya kingdoms in Guatemala and Honduras.

But with no written texts to guide them, archaeologists have long been puzzled about how the people of Teotihuacán lived. By tunneling into the monuments and carefully excavating the smaller structures nearby, they have been able to sketch out some answers. They now envisage a verdant, multiethnic community with rich social strata, including merchants, traders and artisans from across Mexico.

Less clear is what Teotihuacán's politics were. Two schools of thoughts have emerged. One sees a city ruled by a warlike king, unchallenged and infallible, who guided a strict state with an iron fist. The other envisions a mercantile state in which several powerful families vied for influence but could not wrest supreme control and therefore played a careful political game. Now two projects are racing to settle the debate and solve the puzzle. One thing is agreed on, though: in the end, Teotihuacán was not a place where gods were born but a place where men built them out of blood and stone–and a place where they eventually dashed them to the ground.

An Elusive King

Reconstructing the politics of a bygone culture is no easy task. Picture landing in Washington, D.C., 1,400 years after its collapse and trying to understand its people. Did they worship Abraham Lincoln? Were they a military state? Did they conduct elaborate rituals in the reflecting pool? Which of their rulers or priests lived atop the Washington Monument? These are the kinds of questions that Saburo Sugiyama of Aichi Prefectural University in Japan has asked about Teotihuacán for more than 35 years.

Three major structures dominate the site: the Sun Pyramid is the tallest, the Moon Pyramid is the second tallest and the terminus of the Avenue of the Dead, and the smallest, on the same side of the avenue as the Sun Pyramid, is the Temple of the Feathered Serpent. (This last temple is often called Quetzalcoatl, after the similar-looking

Mexica god that was probably inspired by this older version.) Although the Feathered Serpent structure is smaller than the others, many have suggested that it was the most significant and housed potent kings or priests. Located at the very center of the original city, it is made up of two pyramids, surrounded by a squat walled complex.

In 1989 Sugiyama and archaeologist Ruben Cabrera Castro of Mexico's National Institute of Anthropology and History unearthed a crucial new clue to the story of Teotihuacán: 18 bodies near the temple that had been ritually sacrificed and laid out. Soon after, the research duo found more such bodies and then, while tunneling under the temple, still more—well over 100 in all. The individuals they recovered, mostly male, seemed to be warriors from other lands, suggesting a martial society in which rule was enforced at the tip of an obsidian blade.

Sugiyama began to construct a highly specific idea of what kind of place Teotihuacán was in its heyday. He envisioned an ironfisted king who ruled with unquestioned authority in the city. "This city was created, materialized with power. You can't just suggest your ideas by saying, 'Hey, guys, let's make some buildings,'" he says. "You need to convince people with power."

Just outside of Teotihuacán, in a small town, Sugiyama is sitting in his laboratory. The building contains more than two million objects found during the decades since he started excavating. He has a quiet, thoughtful demeanor in person, and it is easy to picture him as the vagabond hippie he was in the 1970s, leaving his native Japan to travel the world. But when it comes to archaeology, few are as ambitious and driven.

In 1998, frustrated that he had not found a tomb or other direct evidence of his kings, Sugiyama decided to tunnel under the giant Moon Pyramid in search of them. It was an ambitious project, but it offered the best way to understand the power structure of the society and was a good place to search for a tomb. Like any big city, Teotihuacán was built in stages. Anthropologists think that starting in 150 B.C., various factions coalesced around this spot in a lush valley and formed an alliance of sorts. They built their city in spurts, first here, then there. Sugiyama's tunneling revealed that the Moon Pyramid was one of the first big structures they created, in a.d. 100.

But they did not erect it all at once. From 1998 to 2004, Sugiyama tunneled past seven previous versions, built one over the other, like Russian dolls. The fourth version—built early in the third century A.D., according to carbon dating—was a major upgrade from the previous one. This was a time of growth and perhaps even the birth of an empire.

Although Sugiyama had yet to locate his kings, his discoveries fit neatly into an emerging picture of Teotihuacán's rise to power. Whether behind a king's banner or not, as Teotihuacán built up its city, it started reaching out its fingers. Some 450 miles southeast of Teotihuacán, in what is now Chiapas, lie the ruins of a small city called Los Horcones that emerged at around the same time. Even this far away from Teotihuacán, Claudia Garcia–Des Lauriers of California State Polytechnic University, Pomona, who has excavated the site, sees its definitive imprint: "The fact that you lay out your main plaza area for your entire community to kind of echo the way that one walks down the Avenue of the Dead and enters into a large open plaza—that's really powerful. You are trying to invoke [Teotihuacán]."

The similarities are subtle but noticeable to a trained eye, Garcia–Des Lauriers says. Given its layout and the style of its pottery, Los Horcones was either a military outpost or a close trading partner with Teotihuacán. Interestingly, the road Garcia–Des Lauriers suspects to be impersonating the Avenue of the Dead does not lead anywhere—it abruptly ends at a large boulder. Yet the dead end makes sense if people who had traveled to the parent city were trying to create a miniature version of it. "This is really just like Mecca in Islamic societies," Sugiyama says of Teotihuacán. "It would have seen pilgrimages."

Similar cities have been discovered in recent years all over Mexico. Scientists now believe that Teotihuacán controlled a region far more vast than any had expected before those ruins came to light—covering much of modern southern Mexico and stretching into Honduras. As its domain expanded, goods such as limestone and feathers poured in from the far reaches of the Mesoamerican world. In fact, in the jungle to the east, controlled by the legendary Maya kings, scientists have found references to a mysterious mountain city to the west, where the people were as thick as reeds in a marsh. Many suspect this western place was Teotihuacán, which apparently even exerted control over the Maya megacity of Tikal in Guatemala. Evidence from another Maya city, Copán in Honduras, suggests that a person from Teotihuacán assassinated its king and established his own rule.

Such an empire, Sugiyama says, needed a strong, charismatic king. He envisions one such leader living around A.D. 219, watching over monumental construction projects and ushering in an empire that would last 500 years.

Eventually, inside the fourth version of the Moon Pyramid, Sugiyama found a stamp of that empire–remains of 12 humans (most likely captives) and more than 50 animals, including wolves, jaguars and eagles, arranged in an elaborate configuration. The layout now looks to be a representation of a creation myth, with a cluster of knives in a circle like a sundial pointing north. "Again, we found very strong evidence of the state–decapitations, knives," he says. "Again, we didn't find the body of a ruler."

In ancient Maya cities, almost every building of note contains a king under the front steps or in the building itself, elegantly preserved and attended by jars of precious spices and stones. Yet to date, not a single king has been discovered in Teotihuacán.

Monarchy or Corporate Society?

Despite the missing king, Sugiyama is still convinced that Teotihuacán was a monarchy, like ancient Egypt, complete with a godlike head and a military apparatus focused on keeping its multiethnic citizens in line. But other experts take a different view. "You cannot have a multiethnic society ruled by one person. They would have had coups d'état all the time," insists archaeologist Linda R. Manzanilla of the National Autonomous University of Mexico. "Teotihuacán had a corporate society, which is the opposite of what the Maya had."

Based on her own excavations, Manzanilla argues that Teotihuacán was ruled not by a supreme leader but four powerful houses, all vying for dominion like something out of *Game of Thrones*. If Sugiyama's version of Teotihuacán is like Egypt, think of Manzanilla's as a little like the Roman Republic—a mighty state ruled by committee. Kings in Teotihuacán were figureheads controlled by the ruling classes from four houses, which were represented in the iconography of Teotihuacán: coyote, feathered serpent, jaguar and eagle. Each house dominated a quadrant of the city and sent representatives to a central governing building with an administrative section for each house. The feathered serpent and jaguar were the strongest houses; thus, the most ornate temples—the Pyramid of the Sun and the Feathered Serpent Temple—were on their side of the Avenue of the Dead.

Both Manzanilla's and Sugiyama's camps say the other's claims are baseless. How can two people study the same site for decades and come back with totally different conclusions? Partly, it may be that Manzanilla looks at Teotihuacán through a very different lens. Where Sugiyama has spent his career looking at the Teotihuacán equivalent of the Washington Monument, Manzanilla says she has been digging up Georgetown: her picture of Teotihuacán stems from her work over the past 20 years in the everyday homes of its people. In the 1990s Manzanilla excavated Oztoyahualco, an artisan apartment compound in the northwest of the city—an area she thinks was controlled by the house of the eagle. Unlike Maya apartment complexes, which have just one shrine, this compound had many different shrines from different traditions. For Manzanilla, this multiethnic quality defines Teotihuacán. She sees a place that could thrive only because wealthy landowners from abroad controlled trade corridors and the goods that fed the community's rapid growth. With so many powerful factions, it would have been very hard for a lone tyrant to hold the population under his thumb. When the city banded together in 150 B.C., Manzanilla says, one group would not have had a monopoly on resources as some of the Maya rulers had. She reasons that the rulers of Teotihuacán relied on taxes from the provinces to build their empire. As such, each shareholder would have leverage in the use of that power. Those with control over the richest territories, such as the Feathered Serpent Temple, had the most leverage over decisions.

This kind of forced power sharing is rare in the ancient world but not unheard of. Rome and Greece were, of course, republics for many years. Likewise Mohenjo Daro, an ancient city in Pakistan's Indus Valley in 2000 B.C., seems to have shared power with a settlement called Harappa, and Tiwanaku in Bolivia shared power with Wari to the north until A.D. 1000.

Sharing with another city you cannot defeat is different than sharing within a city, however. Manzanilla admits that most ancient civilizations had a single ruler and that the setup she proposes is a little odd. She believes, though, that in any given region, some cultures are bound to experiment with joint rule. Yet the strongest evidence she has to support her idea is not something she found but rather something she did not find. Something no one has found.

"Where is that powerful king depicted? Where is he buried? Where is his palace? Can you imagine a site like Teotihuacán, with 125,000 people, [having] a single ruler? His living place and his burial place should have been outstanding. No doubt. And we don't see that," she says. "You'd see him in the vessels, in the throne, in the stelae, in the palaces themselves."

What she does see is a four-petaled flower inscribed throughout the city. Historian Alfredo López Austin of the National Autonomous University of Mexico says that this symbol may represent the four houses that ruled the city. Like Rome, such a place would have been rife with plotting and power plays. As Teotihuacán expanded and its influence spread, the elites acquired increasingly more control. Markets were paved and expensive limestone began appearing throughout the city. The elites of all four houses became greedy and competitive.

To some extent, the data back up this scenario. Increasingly, scientists have been finding that people throughout Mexico moved to the city but maintained their ethnic heritage for hundreds of years. Just as New York City has Spanish Harlem and Chinatown, Teotihuacán had districts for people from Oaxaca to the south, for the Maya and for those from the corridor connecting to the Gulf of Mexico. For example, take the neighborhood of Teopancazco, just south of the Temple of the Feathered Serpent. In the early part of this century, Manzanilla excavated the site, which was dominated by elites from along the trade route to what is now Veracruz, east of Mexico City, and whose symbol she thinks was the feathered serpent. With such a lucrative territory, they would have controlled massive wealth and thus might have sponsored the nearby temple for their feathery emblem.

Manzanilla's work showed that these feathered serpent elites dined on 12 different fish species, salted and smoked, from their home along the Gulf of Mexico, 130 miles away, and decorated their clothing with gulf seashells. "There was a competition between elites to show the best of cosmetics, pigments, hides, cotton clothes, attire, headdresses," she says.

Nowhere was this competition more blatant than in the burials. Since 2005 Manzanilla has painstakingly analyzed several fascinating graves of elite adolescents at Teopancazco. The children are decorated in elaborate fashion, with cinnabar, greenstone and mica from around the empire, plus lots of material imported from their home region.

The wealthy lived in increasingly lavish style for centuries, building up their palaces and lining them with stones that had to be transported from miles away. But the high life could not last forever. In A.D. 350 something snapped. Twenty-nine elites seemed to have been decapitated, and their heads were adorned in a manner only found in the ancient Veracruz region. Manzanilla suspects this was a "termination" ritual, denoting some kind of cultural transformation. At this time, the Feathered Serpent Temple was replaced by one in front of it that displayed jaguars, like an artist redoing a sketch. In fact, after this point, few if any feathered serpents appear anywhere in the city, a sign that the Veracruz people no longer held much sway, in Manzanilla's view.

But they were not gone. After that, a new line of elites must have ascended, and it appears that they continued to prosper in Teopancazco for two more centuries. Then, in A.D. 550, the city of

Teotihuacán burned. No one knows why. Manzanilla says that there is no sign of an invading force. What is in evidence is a large gap between the rich and poor. Analysis of human remains from the site indicates that many of the wealthy were healthy, whereas the poor were malnourished, had back problems from carrying heavy loads and even suffered conditions caused by lack of sunlight—perhaps from slaving away in some workshop. "My guess is that the intermediate elites revolted against the ruling elites," she says. "The ruling elite tried to control this movement too late. These people already had a lot of interests, a lot of alliances in the corridors, and they revolted."

After that, Manzanilla theorizes, a contrite leadership focused its building effort on housing across Teotihuacán and not on elaborate temples. Another century or so more, and the city collapsed for good. For the elites, she says, this sequence of events would have meant simply picking up and returning to their homelands in the corners of the empire.

Teopancazco reveals a slice of life from Teotihuacán separate from the grand temples. The city was diverse, but it was not a melting pot. Instead it was a patchwork, with each culture keeping its identity and operating in fierce competition with its neighbors for prestige and authority. This patchwork may explain why there was no uniform written language or images of kings as there was for the Maya. Such representation would have meant tilting the scales too far in one side's favor.

Perhaps the most fascinating aspect of Teotihuacán archaeology is that both Manzanilla's and Sugiyama's theories could be true. George Cowgill, an emeritus professor at Arizona State University, has proposed that Teotihuacán vacillated between both models, occasionally switching when the right ruler grabbed power. Neither Manzanilla nor Sugiyama likes this compromise, however.

Chamber of Secrets

Archaeologists might never know who ruled Teotihuacán for sure. But very soon Sergio Gómez Chávez may find a crucial piece of the puzzle. Gómez Chávez, an archaeologist at the National Institute of Anthropology and History, has been digging at Teotihuacán for decades, both in common people's homes, such as the city's Oaxaca neighborhood, and in the grand temples. That work included an excavation of the Temple of the Feathered Serpent's drainage system in an effort to restore it to protect the structure from further water damage during storms. It turns out the ancient drainage system works perfectly. But Gómez Chávez found that it was purposefully stuffed up with 50 bodies missing arms and legs. Who builds a drain only to ceremonially plug it? Gómez Chávez knew that the blocking of the drain occurred around the same time that the fourth version of the Moon Temple was built. What if it was done on purpose? What if the city's denizens wanted to flood the area every year, like the reflecting pool in Washington, D.C. (the people of Teotihuacán were known to divert rivers, among other waterworks projects)? It was during one such flooding episode, on a fateful Thursday in October 2003, that the archaeologist made his most shocking discovery.

"I'm coming in to work like any normal day, and [workers] inform me that a big hole has formed" near the temple, Gómez Chávez recalls. He rushed to the site, and sure enough the rain from the night before had opened a perfectly circular tunnel going straight down into blackness. He did not hesitate. He told one of the workers to find a rope. Though not a skilled mountaineer, he managed to tie himself to the rope, and several workers lowered him into the hole, hand over hand. The hole–a foot or two wider than his shoulders–went down some 50 feet and stopped. It was almost like a well except when Gómez Chávez got to the bottom, he found that the dirt and rock on either side was loose, as if it was put there to seal a horizontal tunnel going in both directions. At the top of that fill was a gap that he could almost peer through. "I couldn't sleep for a week, because you don't know what's in it," he recollects.

It seemed that a straight, horizontal tunnel, perpendicular to the shaft, had long ago been filled in with stones and dirt to seal it forever. One end went to a ceremonial entrance behind him, long hidden from view. The other, Gómez Chávez soon learned, continued beyond view–straight to the heart of the Feathered Serpent Temple. The hole he had found was perhaps for ventilation or for light or a view of the stars. Without it, the tunnel may have stayed secret forever. Thus began a 10-year effort to clear the tunnel and find what it hid. The tunnel was filled around that crucial third-century era of expansion. Using radar and sonar, Gómez Chávez learned that it led to a series of three chambers under the center of the pyramid. So he carved his way toward them, one foot at a time. Today he is almost at the end.

A few months ago Gómez Chávez invited me to see the tunnel. Now I am here at his site at the foot of the Temple of the Feathered Serpent, in the wide expanse that he believes was once allowed to flood every year. Walking along the tunnel, I am struck by how chilly it is. It is a hot autumn day outside, but down here it is moist and cool. Gómez Chávez says that when the people of Teotihuacán used this tunnel, the water table was right at their feet. This, he guesses, might have been the reason they put the tunnel at this height. A cold, dark place with water along the ground might have represented the underworld. In several places, I see remnants of a special clay they used to line the tunnel. Peppered with sparkly pyrite, the clay looks almost like twinkling stars. "This is no longer a temple dedicated to Quetzalcoatl but a temple to commemorate the beginning of mythic time," he reflects.

So far Gómez Chávez and his colleagues have removed almost 1,000 tons of filler from the tunnel. In so doing, he has found sealed compartments, one after the other, which he has opened to expose ever more elaborate offerings, including masks, weapons, even reed mats that would have functioned like thrones. "The concentration of materials–it's just amazing. It's unbelievable the things that he is finding," Sugiyama says. Now empty, the compartments appear as pits lining the tunnel floor. Walking on wood planks, I have to be careful not to slip into one.

In the spring of 2013, using a robotic drone, Gómez Chávez reached two of the three major chambers at the end of the tunnel. One contains dozens of quartz spheres, the other pyrite mirrors. No one has any idea what they were or has seen anything like them.

Standing by those chambers, I squat and try to peer ahead at the last chamber, still sealed. It is a few feet lower than where I am. "It's very wet down there," Gómez Chávez says. "With all the mud and water, we have had to go slowly now."

Gómez Chávez imagines this place as an elaborate ritual site, where men left the earthly world for a moment and emerged as kings. Sugiyama, slightly irritated that he did not find the tunnel during his own investigation, says there is no better candidate for a royal tomb. Even Manzanilla admits that the discovery of a royal tomb would be a big deal for Teotihuacán. She insists it will not hurt her theory, but that depends on what is inside.

Meanwhile Manzanilla is excavating her own site, called Xalla, looking for a smoking gun. A series of five structures not far from the Sun Pyramid are arranged in a diamond shape, almost like a four-petaled flower with a shrine in the middle. This, she thinks, is the administrative center of the city, where each of the four houses sent emissaries to look after their business.

Either way—whether Manzanilla proves her theory of distributed power or Gómez Chávez discovers an all-powerful king buried at the end of the tunnel—Teotihuacán will never be the same. At long last, the lives of its people, like ships coming out of the fog, are slowly emerging from 1,300 years of mystery to tell their story.

Referenced

Human Sacrifice, Militarism, and Rulership: Materialization of State Ideology at the Feathered Serpent Pyramid, Teotihuacan. Saburo Sugiyama. Cambridge University Press, 2005.

Corporate Life in Apartment and Barrio Compounds at Teotihuacan, Central Mexico. Linda R. Manzanilla in Domestic Life in *Prehispanic Capitals: A Study of Specialization, Hierarchy, and Ethnicity*. Edited by Linda R. Manzanilla and Claude Chapdelaine. University of Michigan Museum of Anthropology, 2009.

About the Author

Erik Vance is a science writer based in Mexico City. His last article for Scientific American *examined Mexico's struggle to make science pay off.*

A Tale of Two City-States

By Tali Trigg

I'm writing this on a flight from Hong Kong where news has just broken that the father of the Singaporean city-state Lee Kuan Yew has passed away. On the airplane TV, BBC shows a 1993 interview with him, sharing his thoughts on the country's foundation. I didn't catch all the quotes word-for-word, but to paraphrase, he noted that at the time of its independence, Singapore was left on its own, and had to quickly figure out how it could be useful to the world and stand on its own. If not, the city could otherwise support no more than "... 150 fishermen in its natural state," he noted somberly.

Today, Singapore has an outsized presence in trade and finance, with a multicultural population of three million, compared to 150 pirates and fishermen who Sir Thomas Raffles found there upon arrival in 1819. Similarly, Hong Kong (not quite a city-state, but close to it in practice) went from being a sleepy fishing village to now having the 15th highest GDP per capita (PPP). And both are the envy of urban and transport planners worldwide, with well-functioning transit systems plying high-density multi-use corridors.

Neither one is of course really a "city-state" in the old Venetian sense, but they stand as powerful symbols to both older cities struggling to reinvent themselves post-manufacturing as well as countries building new cities from scratch. If a sleepy fishing village can turn into a prosperous nation (much like Iceland went from being a primarily cod-fishing entity of about 143,000 inhabitants in 1950 to a diversified and developed nation today), what can others learn from Singapore and Hong Kong?

Before we proceed, I have to add the requisite caution on needing to take local context into account. What worked in Singapore and Hong Kong doesn't necessarily work elsewhere, of course. For one, they have some fairly unique relationships to their neighbors (Malaysia and China respectively), as well as very strategic maritime locations.

But accepting for differing cultural and historical contexts, what can Singapore and Hong Kong teach other cities, rather than countries?

As an example, let's consider foreign talent. Ask any expat considering a move to the region, and they'll tell you it's often a decision between the two. Singapore is more of a tidy, well-ordered place, where Hong Kong has more of a vibrant feel with an incredible food scene. Then again, it's a world where diversity thrives, so why choose?

However, it's not really about choosing between the two, it's that the two are considered so attractive in the first place; something the two have achieved through a holistic and integrated focus on urban and transport planning. In fact, Hong Kong's transit agency (MTR) is now world famous and operates transit systems elsewhere, from Melbourne to Stockholm. And in Singapore, they've managed to turn their national airline and airport into a showcase of the country's qualities, with Singapore Airlines being voted the number one airline on the planet, and the same accolade going towards its home airport (Changi). So does it come down to infrastructure alone?

Maybe. Yet, while there is no one thing the two "city-states" have achieved, it's rather how they've turned their strategic locations into their advantage, becoming first-rate maritime and financial hubs thanks to excellent underlying infrastructure and a clear legal framework, making it easy to do business. To achieve all this, especially in the service sector, these two places needed to attract the relevant talent by building the necessary infrastructure, all with an eye towards preserving some of the original green spaces.

This was especially important in places where space is at an incredible premium. The result? Lots of building upward, with very good density, and thus excellent utilization of public transit networks. All in all, it makes for a good start, but are we missing something as an explainer?

In a recent *Forecast* magazine article (an offshoot of globalist *Monocle* magazine), the former and current capitals of Kazakhstan were compared, mostly unfavorably to the new capital, for lacking a "soul." While *Forecast* bemoans the usage of such simplistic description, I wonder if there isn't something to it.

Similarly, in the same magazine, the United Arab Emirates' eco-city-from-scratch (Masdar City) is described as a sluggish project that might well turn out to be a white elephant, despite the earnest intentions of the foreign architects given free reigns and ample pay. By next 2016, the goals were to house 50,000 full-time residents. Currently, only about 300 students live there. Perhaps they really are missing that "soul" or a *je ne sais quoi*?

As these new cities take shape, and older ones reinvent themselves (to varying degrees of success), Singapore and Hong Kong teach us that an underlying integrated land-use and transport planning will serve cities well—they not only improve the lives of those there, but also increase their competitiveness in attracting a high-value pool of foreign talent.

The views expressed are those of the author(s) and are not necessarily those of Scientific American.

About the Author

Tali Trigg is an energy analyst, technology policy advisor and writer. His work includes research and analysis on energy and transportation, with an emphasis on the role of cities in shaping transport energy demand and mobility solutions. His blog covers the wide range of mobility and energy, with deep-dives into numbers and maps, but is keen to cover anything transport-related and under-reported. Opinions are his own.

The Rise of *Homo verticalis*

By Roland Bouffanais and Sun Sun Lim

The evolution of humankind is intimately intertwined with the development of our living spaces. All forecasts point to an almost complete urbanization of our planet by 2100. This process was accelerated by the industrial revolution, leading to rapid urbanization which has now entered a truly explosive phase. Predictions about our urban futures are often associated with dystopian depictions of cities being bogged down by massive pollution, traffic congestion and overcrowding.

However, one should not forget that cities have always been places of intense intellectual burgeoning and social creativity. Indeed, current utopian visions of smart cities envisage next-generation urban spaces as sustainable metropolises that facilitate social connections, stimulate entrepreneurship and liberate human mobility. The 19th-century French poet Charles Baudelaire described the *flâneur*, the artist-poet of the modern metropolis: someone who wanders about a city with no particular purpose, enjoying consummate freedom while exploring its vast expanse.

Beyond such romantic notions of individual mobility, what is the true extent of the urban citizen's freedom of movement? Over the last decade, advancements in complexity science have allowed us to dissect, with unprecedented accuracy, the fine details of human mobility in dense urban areas, thus enabling more strategic planning of our built environment. This breakthrough was triggered by the smartphone revolution, which made it possible to track and record the movements of millions of *Homo sapiens*.

Mining this "big data" revealed the burstiness of human wandering–short time frames of intense activity followed by longer periods of slow activity–that necessitated particular responses in urban planning. These studies, albeit fascinating and groundbreaking, were mostly carried out on large tribes of North American or European *Homo sapiens*, who live and wander in mostly flat cities that strain at their edges, congealing in an urban sprawl. Notably, therefore, the admittedly illuminating data were fundamentally limited to horizontal movements.

Yet the Asian urban landscape is quite distinct, manifesting a sprawl of an upward nature. The ongoing intensive urbanization in Asia is distinguished by the emergence of a specific type of city, namely the vertical one. Indeed, verticality is considered as one effective means to cope with high human density, in particular to overcome the known inefficiencies in public transportation associated with urban sprawls. How can we best leverage verticality to build more hospitable and livable cities as urbanization aggressively pushes forth?

While a flurry of brand-new skyscrapers is being erected every day around the globe, we know surprisingly little about vertical human mobility. This stands in stark contrast to our depth of knowledge and understanding of horizontal human mobility. Residents of these vertical cities wander about in three dimensions–up, down and across. Our everyday movements comprise a succession of jumps up and down, traversing floors via stairs and escalators, lubricated by elevator rides from high above the ground to several levels of basements underground.

This vertical mobility allows us to reimagine and reconfigure urban spaces. For instance, it is now common to find a hotel lobby on the 40th floor of a building in Tokyo, a garden embedded in the façade of a tower in Singapore, tennis courts on the rooftop of a sports center in Seoul, and so on. These are the possibilities, and realities, of life in vertical cities for hundreds of millions of Asian urban dwellers. This large tribe of *Homo sapiens* can be referred to as *Homo verticalis*.

How freely and how well can *Homo verticalis* wander given the nature of the vertical cityscape? We should first contrast the basic topographical features of a flat city with that of the vertical one. A bird's-eye view of the flat city would resemble a maze-like structure offering tremendous wandering opportunities characterized by a multitude of continuous paths between any two points.

However, anyone planning to go from the observation deck of a skyscraper to the rooftop bar on a neighboring building will inevitably have to descend to the ground floor and cross several blocks, before ascending again. The maze would thus become multilayered, with a mind-boggling array of possible routes that extend to and across every level of the tallest building in the city.

While the potential for wandering through the vertical city seems limitless, *Homo verticalis* faces many barriers in this man-made environment. In many new buildings, going up or down even a single level necessitates an elevator ride because staircases are tucked away from main thoroughfares. In shopping malls, escalators are strategically positioned to constrain wandering, with patently commercial motives. Similarly, underground train stations that serve millions of commuters daily are designed to both promote and impede vertical mobility, seeking to prevent overcrowding at stations during peak hours, while enabling efficient evacuation during emergencies.

To uncloak the hidden barriers and identify facilitators to movement in the vertical city, designers and urban planners should work closely with complexity scientists to map the hitherto unknown patterns of vertical human mobility. The advent of affordable and efficient sensors to track such vertical movements can help to generate the big data that will shed critical light on these patterns. This new knowledge would be pivotal to the establishment of novel design principles that can enhance mobility and therefore livability in the vertical urban world.

About the Authors

Roland Bouffanais, Ph.D. is associate professor at the Singapore University of Technology and Design. He authored Design and Control of Swarm Dynamics *(Springer, 2016). He was awarded the IBM Research Prize in Computational Sciences (2008) and the ERCOFTAC Da Vinci Award Silver Medal (2007).*

Sun Sun Lim is professor of communication and technology and dean of humanities, arts and social Sciences at the Singapore University of Technology and Design. She recently authored Transcendent Parenting: Raising Children in the Digital Age *(Oxford University Press, 2020). She serves on the boards of the Social Science Research Council, Singapore Environment Council and Media Literacy Council. During 2018–2020 she served as nominated member of the 13th Parliament of Singapore.*

Cities Are Surprisingly Fragile

By Michael Hunter, J. David Selby, and Kevin C. DeSouza

A recent report from the Economist Intelligence Unit highlights the increasing fragility that urbanization can bring. Jakarta's rising urban population is facing challenges due to its unique water infrastructure. Flooding and continued pollution from deep water pumps have caused billions of US dollars in damages and considerable loss of life, especially among lower socio-economic families in flood prone areas. Furthermore, a rise in the population of Bangkok (now the ninth largest population in east Asia) has seen the city's traffic congestion rating rise from 30th to 12th globally in just two years, resulting in the average driver experiencing roughly 64 hours of traffic congestion each year.

Globalization has accelerated the rate of industrialization within east Asian countries and while it has brought economic and social opportunities to these countries, it has also had numerous unintended consequences associate. Alongside the rise of globalization has been an increase in urbanization; 55 percent of the world's population now live in urban areas. A recent World Bank Report discusses 200 million people moving to urban areas in East Asia alone, equivocal to that of the sixth largest population for any single country.

Consequently, the bulk of the economic boons of globalization are being captured by cities. Currently, Tokyo contributes 1.6 trillion dollars to Japan's GDP (almost double the entire nation of Malaysia). Cities are becoming increasingly more complex due to the dual forces of globalization and urbanization, which makes city governance harder, thus social compacts are more likely to go unfulfilled. An example of this is the consistent trend for governments to prioritize continued economic development over sustainable living practices for citizens. In other words, cities are becoming increasingly fragile.

Fragility occurs when there is some break down in how society ought to work. Increased congestion, rising housing prices, rising cost of living, and increased levels of homelessness are all examples of urbanizations increasing fragility. Urbanization and globalization are creating fragility in cities faster than the traditional model of reactive governance, addressing issues as they occur, can address.

We encourage policy makers to think about urban fragility along the same vein as resilience, in fact fragile cities tend to be more damaged by natural disasters than relatively un-fragile cities. By understanding urban fragility, we instead argue for a framework of considering issues in terms of failed social compacts, which should allow policy makers to take a more proactive and holistic approach to addressing fragility.

Social Compacts

Social Compacts are the norms, traditions, and laws that bind society together, through both implied and explicit agreements between government and individuals (e.g., cultural norms). They determine which problems governments should address, and what are acceptable solutions to those problems.

Since cities do not have unlimited resources, they cannot address all compacts. This means that cities will always have some form of fragility, and that compacts have to be prioritized in order to minimize the fragility a city faces.

Therefore, compacts need to be categorized according to a hierarchy of social compacts. Broadly speaking, there are four levels of compacts: fundamental, stability, integrity and prosperity. Starting with fundamental, cities need to prioritize fulfilling these compacts.

Globalization Within the Asia Region

Countries of the Asian region have experienced a tremendous amount of economic growth and development. Economies such as Singapore, Hong Kong and South Korea have all utilized nationwide industrialization and neo-liberal approaches to economic policy in order to provide attractive centers for foreign investment.

In doing so, each country has seen remarkable progress in a short amount of time, with South Korea placing in the top 10 exporters globally, Singapore experiencing double digit annual growth of GDP during the 1970s, and Hong Kong property values increasing by 8 percent per annum for the previous 25 years.

Unfortunately, economic growth associated with globalization has corresponded with a rise in the complexity of governing urban areas. Generally, city governments tend to prioritize the prosperity compact over the concerns of citizens. A top down approach is adopted across the social compact framework, wherein prosperity is focused, integrity is required to uphold the prosperity, and fundamental compacts are merely reacted to as issues arise.

The following east-Asian examples demonstrate issues that have arisen from prioritizing prosperity to the exclusion of other compacts. We suggest that with the correct application of the social compact framework, governments can design better policy to proactively mitigate these problems.

Singapore

Singapore's rapid economic development during the previous 50 years of independence has been recognized globally as a highly successful means of city and state governance. Early on, the government spearheaded a focus on education and industrialization supported by social policy built on meritocratic and neo-liberal values.

Perhaps the most successful and well recognized of these policies, is their Public Housing Scheme. This particular scheme is responsible for providing 90 percent of Singapore's population with home ownership, and is such a successful method that during the 1990s the United Nations would invite other countries to study the policy.

By providing citizens with a scheme that meets their fundamental compact (housing), citizens are better situated to support the prosperity compacts allowing for more sustainable economic growth. A differentiator from other Asian regions, this public housing scheme plays a largely supportive role in Singapore's ability to develop economically. The government's control of housing allows Singapore to regulate and control the housing market to ensure it aligns with the merit-based social values they promote.

Singapore's housing policy was initially successful because it promoted fundamental compacts along-side integrity compacts. The government tied membership of the Public Housing Scheme with conformity to social values. The policy started causing fragility when globalization brought diversification that did not align with traditional social values such as housing for single parents.

The government of Singapore has been slow to respond to this shift because they believed adhering to social values (integrity compact) was responsible for the economic success of the city, instead of meeting all residents' fundamental compacts.

This failure to understand the social compact framework has resulted in a rise in homelessness and an increased stress on the interim and public housing scheme programs. These previously successful programs have increased 400 percent in their application processing times, leaving applicants potentially homeless for up to two years.

Hong Kong

Hong Kong utilizes both public housing, and government leased housing policies to develop itself economically and attempt to regulate the housing market. Over the previous 40 years, the Hong Kong Housing Authority (HKHA) has built over 660,000 rental flats, and is considered one of the largest public housing bodies in the world. Initially, Hong Kong introduced the Home Ownership Scheme as a way to provide low-income households with an opportunity to afford home ownership. Whilst extremely beneficial to Hong Kong's lower income citizens, this scheme also provided massive economic benefits as it allowed the government to relocate labor to aid in industrialization. Between public housing and government owned property, Hong Kong policy was able to link a critical mass of people with industry, which consequently improved both the business and public environments.

However, post the Asian financial crisis, Hong Kong has struggled to offer as consistent of a policy as Singapore. During 2002, the government renounced the development of any Home Ownership Scheme flats and in 2004, the government began selling Home Ownership Scheme projects to private developers. The sale of the scheme's development plans to private enterprise counters the exact purpose of the scheme's initial implementation–to provide subsidized housing to those who could not afford to enter the private market. In the following years, as low-income households were left with unfulfilled fundamental compacts, the private real estate market prices surged over 275 percent.

Currently, the Hong Kong government is only responsible for roughly 40 percent of the housing market, which is largely limiting on their ability to align housing policy with economic development.

The Hong Kong government's decision to prioritize economic prosperity over fundamental housing has resulted in a reactive policy cycle where continued economic development is no longer led by the industrialization and mobilization of a workforce, and instead is decided by private developers purchasing power.

The property market inflation in Hong Kong provides a perfect example of the increasing fragility that urbanization can bring to cities. Almost a decade since the Home Ownership Scheme has been re-implemented, homelessness figures are still increasing roughly 20 percent per annum with a record 1.37 million citizens living below the poverty line. These figures are yet to include the numerous other citizens who utilize 24-hour McDonalds and internet cafés for shelter, of which are estimated at increasing roughly 50 percent in the previous 12 months. It is crucial that governments consider the fundamental level of social compacts before aligning themselves with economic development.

Seoul

Seoul largely focuses on continuous urban renewal as a way to attract both foreign tourism and investment and has previously found economic success using this method, utilizing urban transformation as a way to attract foreign investment and compete globally.

Similar to the previous examples, Seoul's economic development is largely due to its ability to meet the rising needs of globalization and benefit from industrialization. However, the city goes one step further in utilizing urban transformation as a way to physically promote an innovative image, advertising its transformation of fundamental compacts as a means to attract tourism and foreign investment. The Cheonggyecheon restoration project was initially globally successful as an attractive solution to rising urbanization issues, providing further economic interest from business and tourists.

Unfortunately, aligning urban transformation with economic goals and a global outlook has prioritized the attractiveness of foreign investment over the fundamental quality of life provided to all citizens. Specifically, whilst foreign investment can have positive effects on economic growth, it can also outsource the conversation of urban transformation to external interests, negatively effecting the democratization of the conversation for local citizens.

Furthermore, the South Korean federal government has recently re-focused their attention on welfare, demonstrating that whilst locally urban transformation provides economic benefits, it has numerous consequences that if not initially considered will undermine its initial positive economic capabilities and create a reactive cycle of policy.

Proactive Policy for Fragility Mitigation

In the face of continuing globalization and rising urbanization, it is increasingly attractive for cities to prioritize the higher-level compacts. The examples mentioned highlight the numerous issues that arise from prioritizing prosperity and integrity compacts ahead of the fundamental needs of citizens. Importantly, it demonstrates the reactive policy cycle that occurs when governments fail to consider the hierarchy of social compacts.

Using this Social Compact framework, we can understand cities as a continuing process, of which agreements between citizens and state must be continually balanced. Cities are evolving, dynamic areas and understanding their fragility allows us to maintain a continued discussion capable of adapting as the city evolves. Rather than reacting in the face of city crises, we utilize this framework to understand cities' evolving dialogue and adapt accordingly.

The views expressed are those of the author(s) and are not necessarily those of Scientific American.

About the Authors

Michael Hunter is a researcher at the QUT Business School, Queensland University of Technology.

J. David Selby is a Public Policy and Public Administration PhD candidate at Arizona State University.

Kevin C. DeSouza, PhD, is a Professor of Business, Technology and Strategy in the School of Management at the QUT Business School, Queensland University of Technology.

Section 2: The Science of Cities

Bigger Cities Do More with Less

By Luís M. A. Bettencourt and Geoffrey B. West

For centuries, people have painted cities as unnatural human conglomerations, blighted by pathologies such as public health crises, aggression and exorbitant costs of living. Why, then, do people throughout the world keep leaving the countryside for the town? Recent research that is forming a multidisciplinary science of cities is beginning to reveal the answer: cities concentrate, accelerate, and diversify social and economic activity.

The numbers show that urban dwellers produce more inventions and create more opportunities for economic growth. Often large cities are also the greenest places on the planet because people living in denser habitats typically have smaller energy footprints, require less infrastructure and consume less of the world's resources per capita. Compared with suburban or rural areas, cities do more with less. And the bigger cities get, the more productive and efficient they tend to become.

The Power of Population

This new, more quantitative science of cities is becoming possible because of the increasing availability of information–official statistics as well as novel measures of human and social activity–on cities and metropolitan areas worldwide.

By sifting through this flood of data, covering thousands of cities around the world, we have unveiled several mathematical "laws" that explain how concentrating people in one place affects economic activity, return on infrastructure investment and social vitality. Despite the rich diversity of metropolitan regions across the U.S., China, Brazil and other nations, we found a remarkable universality in the way that socioeconomic characteristics increase with a city's population. For example, if the population of a city is doubled, whether from 40,000 to 80,000 or from four million to eight million, we systematically see an average increase of around 15 percent in measures such as wages and patents produced per capita. If eight million people all live in one city, their economic output will typically be about 15 percent greater than if the same eight million people lived in two cities of half the size. We call this effect "superlinear scaling": the socioeconomic properties of cities increase faster than a direct (or linear) relation to their population would predict.

The data also reveal that cities' use of resources follows a similar, though inverted, law. When the size of a city doubles, its material infrastructure–anything from the number of gas stations to the total length of its pipes, roads or electrical wires–does not. Instead these quantities rise more slowly than population size: a city of eight million typically needs 15 percent less of the same infrastructure than do two cities of four million each. This pattern is referred to as sublinear scaling. On average, the bigger the city, the more efficient its use of infrastructure, leading to important savings in materials, energy and emissions.

Our findings also show that these patterns of increased productivity and decreased costs hold true across nations with very different levels of development, technology and wealth. Although we have much more information for cities in richer parts of the world, we are beginning to obtain good data from rapidly developing countries as well, and they seem to fit the same mold. The gross domestic product for cities in Brazil and China, for instance, closely follows the same superlinear curve that western European and North American cities exhibit, though starting from a lower baseline. We believe that the pattern holds true because the same basic social and economic processes are at work, whether in São Paulo's favelas, under Beijing's smog-filled skies or along Copenhagen's tidy streets.

Although urban superlinear scaling, which represents the average, idealized behavior of a city of a given size, prevails around the globe, actual cities deviate to varying degrees from the roughly 15 percent enhancements that come with size. Detailed data covering 40 years show, for example, that San Francisco and Boston are richer than their size would indicate, whereas Phoenix or Riverside, Calif., are somewhat poorer. Curiously, these deviations persist for decades: cities tend to stay remarkably close to their overperforming or underperforming histories. For example, cities that have attempted to improve their lot by creating conditions for the "next Silicon Valley" have often had disappointing results. Our research suggests that certain intangible qualities of social dynamics—more than the development of material infrastructure—hold the key to generating virtuous cycles of innovation and creation of wealth. These processes, such as the development of a spirit of local entrepreneurship, a reputation for cutting-edge novelty, and a culture of excellence and competitiveness, are difficult to design through policy because they rely on the dynamics of a city's social fabric across many dimensions. We expect the results of this exciting area of research will lead to better "recipes" for sustainable socioeconomic development.

What we can say with certainty, however, is that increased population promotes more intense and frequent social interactions, occurrences that correlate with higher rates of productivity and innovation, as well as economic pressures that weed out inefficiencies. In a city with high rents, only activities that add substantial value can be profitable. These economic pressures push urbanites to come up with new forms of organizations, products and services that carry more value added. In turn, higher profitability, excellence and choice tend to attract more talent to the city, pushing rents higher still, fueling the need to find yet more productive activities. This feedback mechanism, in a nutshell, is the principal reason cities accelerate innovation, while diversifying and intensifying social and economic activity.

Denser But Greener

Although cities create economic opportunities in rich and poor countries alike, people living in wealthier areas find it difficult to imagine why so many inhabitants of poor countries are attracted to places such as Nairobi, Lagos or Mumbai, where newcomers often end up in slums marked by pollution, crime and disease. These appalling conditions, however, should remind residents in developed nations of their own urban past. When Charles Dickens wrote about life in mid-1800s London or when Jacob Riis photographed the Bowery district of New York City's Lower East Side in the late 1800s, each was reporting similar circumstances. These cities grew explosively during the 19th century—sevenfold for London and almost 60-fold for New York. Well-run modern cities have demonstrated that pervasive ills are not inescapable. The problems result primarily from nonexistent or poor planning and a lack of good governance. The development of these organizational traits may, in fact, be the most important and long-lasting effect of urbanization because it paves the way for socioeconomic development at the national level.

Some benefits besides wealth and innovation come about even when not legislated. One notable example is the impact of cities on the environment. Quality data are only now beginning to emerge, but we can already see that the largest U.S. cities have the lowest carbon dioxide emissions per capita. This gain is mostly an unplanned by-product of people living at greater densities because the bulk of the savings comes from energy-efficient public transportation and simple walking instead of driving, which is almost 10 times more energy-intensive.

Environmental efficiency becomes more challenging for developing nations such as India or China, where much urban infrastructure still needs to be built, although the trade-offs between a need for rapid growth versus the steps to achieve clean growth remain poorly understood. Still, urbanization may ultimately remain the most sustainable solution to our planet's environmental challenges.

Unbridled growth can nonetheless create crises that, in the extreme, could cause a city to collapse unless major innovations are found to stimulate new cycles of growth. In this sense, cities are never in a state of stable equilibrium. They exist in a dynamic balance—a kind of tug-of-war—between the forces that bind them together and those that can potentially tear them apart. That tension is another reason cities drive innovation: many of civilization's greatest inventions have come from dire necessities. Just think of plumbing, electricity and even democracy—not to mention coffee shops.

The ongoing challenge for urban growth is whether human creativity can keep innovating sufficiently fast to sustain ever expanding urban populations while decreasing our per capita consumption of resources and impact on the planet. As long as this trend can continue, cities will grow ever larger and will be the inevitable future of a more creative and prosperous humanity.

About the Authors

Luís M. A. Bettencourt and Geoffrey B. West are both theoretical physicists at the Santa Fe Institute and Los Alamos National Laboratory who have introduced methods from mathematics and physics into biology and the social sciences.

Grid Unlocked: How Street Networks Evolve as Cities Grow

By Sarah Fecht

The world's cities are absorbing one million additional people every week—and by 2030, they could consume an extra 1.5 million square kilometers of land, or roughly the area of France, Germany and Spain combined. What would be the best ways for those cities to grow? A new study examines how—before urban planners existed—a group of Italian villages evolved into suburbs outside Milan today. Such studies may eventually help planners optimize future developments.

"We know few things about how cities grow naturally," says Emanuele Strano, a doctoral candidate studying urban geography at Switzerland's École Polytechnique Fédérale de Lausanne who authored the study. "Urban planners believe that with regulations we can control the growth of cities. The question is, how can we control a thing if we don't really know how it behaves?"

The new study takes a step toward that essential understanding. Strano and his colleagues—a group of computer scientists, mathematicians, physicists and urban scholars—teamed up to provide the first quantitative analysis of how unplanned street networks evolve over time. Their results were published March 1 in *Nature*'s "Scientific Reports." (*Scientific American* is part of Nature Publishing Group.)

"Historically, a lot of these analyses have been done using not such a substantial quantitative basis," says Stephen Marshall, an urban theorist at University College London (U.C.L.) who was not part of the study. "In the areas that I'm familiar with, people might produce a map of a city, and compare and contrast two maps on a more qualitative basis." Marshall also says that this work marks the first time anyone has looked at how those properties change over time.

Strano and his colleagues studied the growth of Groane, an area outside Milan, Italy. Using maps dating as far back as 1833, the researchers modeled the urban sprawl process as Groane developed from agricultural land into a residential and industrial area, and then into a postindustrial suburb of the Milanese metropolis. At seven time points between 1833 and 2007 the researchers used geospatial data and programming script to construct digital maps (shown above) that treat the street system as a network of links (streets) and nodes (intersections).

Network analyses performed on the digitized map revealed that although Groane had never been subjected to large-scale planning efforts, the street network evolved to be more uniform in size and density, and its predominant shape shifted from triangular to rectangular. Rectangular street networks may arise organically, the authors say, simply because they offer more efficient pathways and allow for more organized subdivision of land.

The study also identified two processes of urban growth: one of exploration, where new streets extend into rural areas, and one of densification, where new streets fill in gaps within the existing network. Co-author Marc Barthelemy, a theoretical physicist at the French Commissariat à l'Énergie Atomique explains that although the group has identified some rules that govern the densification process (which tends to make the grid more orderly and square), exploration is less predictable. He says that his team's future studies will take a closer look at the exploratory process to learn what drives urban sprawl. "We ought to understand it because it might give some hints on how to slow it down."

Whereas modes of transportation changed drastically between 1833 and 2007, the original road layouts remained mostly intact. In fact, the researchers found a tendency for older streets to be the more central ones in modern times–90 percent of Groane's most interconnected roads in 2007 had already existed in 1833. Because streets with greater connectivity tend to have higher traffic flows, Barthelemy says urban planners could compute the connectivity on existing or proposed street networks to understand how people will use the space. (Some planners–such as Space Syntax in London and the University of Strathclyde's Urban Design Studies Unit, of which Strano is a member–are already measuring connectivity to optimize urban designs.)

Planners had already suspected many of the study's findings, but the research provides empirical support for the theories and suggests that natural rules may guide the self-organization of cities. "My suspicion is that some of these rules are working everywhere, because cities are always man-made objects," Strano says.

Marshall cautions that, "to get a complete picture, we need to combine this kind of network analysis approach with an understanding of what's happening on the ground–What are the human actions?" Marshall suggests that factors such as cultural design preferences, the shapes of buildings and small-scale planning efforts may have subtly influenced Groane's street network. Growth patterns also may differ in city centers, rather than at the city's edge, where Groane is located. The authors of the study agree, and they say that answers await further studies of street network evolution in cities of various sizes and geographic locations as well as with differing economic histories.

The implications are important: Correlational studies have suggested that street patterns may impact local economies, physical activity levels, public transportation use, crime rates and social inequality.

"Cities are immensely complex structures," says Michael Batty, a spatial analyst at U.C.L. "When you interfere with a complex system, you get all sorts of side effects." He says that quantitative research such as Strano's could potentially guide planners into enhancing organic growth, rather than stopping it in its tracks–thus helping planners to build better cities and prevent uncontrolled sprawl.

Jane Jacobs: Is There Good Science behind Urban Planning?

By Sarah Fecht

Editor's note: In the 1960s, activist Jane Jacobs (who would have turned 100 on May 4, 2016) questioned whether massive urban planning projects work, and objected to demolishing old neighborhoods for sterile new buildings and highways. She described factors including mixed-age buildings and short blocks as leading to better neighborhoods. Has the state of city planning science changed since then?

In 1961 urbanist Jane Jacobs didn't pull any punches when she called city planning a pseudoscience. "Years of learning and a plethora of subtle and complicated dogma have arisen on a foundation of nonsense," she wrote in *The Death and Life of Great American Cities*. Fifty years later the field is still plagued by unscientific thought, according to urban theorist Stephen Marshall of University College London. In a recent paper in *Urban Design International,* Marshall restated Jacobs's observation that urban design theory is pseudoscientific and called for a more scientific framework for the field.

Although urban design theory is unscientific, Marshall wrote, it is not because the ideas are based on nonsense—many of the classic urban thinkers used observations and small pilot studies to describe how cities work. Jane Jacobs, for example, proposed that a city needs four ingredients to be exuberant: mixed uses, short blocks, buildings that vary in age and condition, and a dense concentration of people. "At the core of this book is a four-part hypothesis that is demanding to be tested," Marshall says. "But when I went to look to see if it had been tested, there was virtually nothing." The problem with urban design, he adds, is that its theories are untested, yet accepted as fact. Marshall proposes to overhaul the way that urban design incorporates science into its fabric, calling for more and better urban science, and for the theories to be challenged with alternative hypotheses and rigorously tested.

Some researchers are already studying cities in scientifically valid ways, but much of this work is being done by physicists and mathematicians who have little use for urban design theory.

"In urban planning, we're like physicists without a particle theory or doctors without a germ theory," says Michael Mehaffy, an urban designer at Portland, Ore.–based design company Structura Naturalis. "We don't have a unifying idea about the nature of what we're looking on. We say we're artists but it's as if we're medieval doctors with potions. . . . We need to recognize that we have a responsibility to use models that are more likely to produce better outcomes."

Urban designers traditionally have doubted the role that science can play in describing or predicting or fixing a city. They assert that cities have an emergent complexity that results from the interactions among people as well as between people and the environment, and that there is an element of human behavior that cannot be reduced to an equation. "I'm not really sure that the city can be studied and predicted any better than we can predict the weather," says Peter Laurence, an architect at Clemson University. "And cities are more complex than the weather."

Geoffrey West, a physicist at the Santa Fe Institute who studies urban growth, agrees. People who study cities "are dealing with maybe one of the most complex systems in the universe," he says. "The idea that you could reduce it to an equation is extremely hubristic." Yet he and others are using computer simulations, modeling and mathematics to begin to tease out the simple rules that give rise to urban complexity and diversity. For example, West has devised formulas that use a city's population to predict the size of the metro police force, the number of people with AIDs and the annual tonnage of carbon dioxide emissions. The predictions are accurate 85 percent of the time. That consistency is guided by universal social dynamics that govern the way people interact with each other and with the environment, West says. The other 15 percent is individual variation between cities.

Although West says he'd like to see urban design's classic theories put to the test and integrated into a scientific framework, he finds his own scientific work as complementary to traditional ways of studying cities. "The thing that makes urban planning so difficult is that the role of design has to be there. That's critical, and it involves things that are outside of science in the traditional sense."

How can these two viewpoints–of science and design–be reconciled? Mehaffy suggests that urban design theory and urban design practice could have a relationship like that of life science research and medicine. A doctor doesn't spend all of his time in a research lab, but he relies on scientific knowledge to guide his decisions on a case-by-case basis. The art comes in the form of tailoring diagnoses and prescriptions for each individual patient.

Marshall says that if urban designers don't build themselves a more scientific foundation, then outside researchers will do it for them. To survive, the field needs to incorporate scientific training into its educational curricula, and cultivate "a concern for testing and validation, critical assimilation of scientific findings from disparate sources, and dissemination of the most reliable, up-to-date findings."

Will the next 50 years be any different from the decades that have passed since Jane Jacobs first cast urban design into pseudoscientific limbo? West thinks that as cities grow and researchers continue to elucidate the influence of design on factors such as carbon dioxide emissions, physical activity and quality of life, there will be an inevitable shift toward scientific thinking. "The city is [the] most important problem we have to face," he says. "What are the underlying principles of urbanization, and what is urbanization going to do to the planet?"

How Traffic Jams Decentralize Cities

By Sarah Fecht

Most of the cities that exist today originally grew from an important marketplace or town square. Over time, they developed multiple centers where people could go to work, shop and play—in fact, no major world city has just a single center anymore. So, what governs the transition of a city from monocentric to polycentric? Economists have suggested polycentrism is driven by business agglomeration—the idea that companies are more successful when they are clustered. But physicist Marc Barthelemy of the Institute for Theoretical Physics in France says that is only part of the story.

In a new paper in *Physical Review Letters*, Barthelemy and his colleague, Remi Louf, have constructed a mathematical model to explain how cities and their surrounding suburbs evolve to be polycentric. Their findings suggest that population size and automobile traffic congestion play large roles in driving the creation of alternative hot spots, even in small- to medium-size cities. "It's an interplay between how attractive the place is, and how much time it takes to go there," he says. At first everyone goes to the city center, but as the city becomes increasingly crowded it becomes more difficult to get there. Eventually subcenters spring up toward the city's outskirts, providing more convenient locations for residents to work and shop. Cities with accommodating transportation networks remain centralized longer, but once population density passes a certain threshold, cities inevitably become polycentric, Barthelemy says.

It's a simple model—for instance, it does not account for zoning restrictions or for the fact that some cities, such as London, emerged as polycentric entities hundreds of years ago when several smaller villages consolidated—but it is still a useful model for exploring polycentricity, says Michael Batty, an urban planner at University College London, who wasn't involved in the study. "Although it is a major simplification of the real world, it has enough elements to make it work, and the dynamics they produce are interesting and plausible."

David Levinson, a transportation engineer at the University of Minnesota, says it is not altogether surprising to find a relationship between population size and the number of urban subcenters. A group of economists made that assumption a few decades ago. Barthelemy counters that the economic models were "fuzzy" and untested. "After 20 pages of calculation, they don't have a prediction and they don't test their model," Barthelemy says. "We can test our results against data."

To test the prediction that the number of subcenters scales with population, the researchers used employment data across 9,000 U.S. cities and towns. "If you know where people are working, you can see these activity centers," Barthelemy says. The data showed that as a city's population increases, so does the number of subcenters. But the two do not scale linearly–if a population grows to 10 times its original size, the number of centers is multiplied by five, on average. Barthelemy says he expects the pattern to hold across cities in other countries of the world as well.

Having a clearer understanding of the evolution of metropolitan polycentricity could prove useful, Levinson says, especially considering that two thirds of the world's population is expected to be urban by the year 2050. "There's a lot of urbanization left to happen," Levinson says. "If planners imagine a city to take a particular form, but that's not the way the city wants to behave, we'll be making unwise investments."

The authors are still fine-tuning their model. Future versions will likely include other modes of transportation, such as public transportation, which would be expected to slow the appearance of new centers of activity.

Barthelemy says the model could also come in handy for estimating traffic delays, gas consumption and carbon dioxide emissions, so that urban planners and policy makers can better understand how those different factors evolve as a city grows. "I think that this opens up the path to some really quantitative insights about cities," he says. "We can take simple mechanisms, simple ingredients, and in the end predict how important properties are scaling with population."

Simple Mathematical Law Predicts Movement in Cities around the World

By Viviane Callier

The people who happen to be in a city center at any given moment may seem like a random collection of individuals. But new research featuring a simple mathematical law shows that urban travel patterns worldwide are, in fact, remarkably predictable regardless of location—an insight that could enhance models of disease spread and help to optimize city planning.

Studying anonymized cell-phone data, researchers discovered what is known as an inverse square relation between the number of people in a given urban location and the distance they traveled to get there, as well as how frequently they made the trip. It may seem intuitive that people visit nearby locations frequently and distant ones less so, but the newly discovered relation puts the concept into specific numerical terms. It accurately predicts, for instance, that the number of people coming from two kilometers away five times per week will be the same as the number coming from five kilometers twice a week. The researchers' new visitation law, and a versatile model of individuals' movements within cities based on it, was reported in *Nature*.

"This is a super striking, robust result," says Laura Alessandretti, a computational social scientist at the Technical University of Denmark, who was not involved in the study but co-wrote an accompanying commentary. "We tend to think that there are lots of contextual aspects that affect the way we move, such as the transportation system, the morphology of a given place, and socioeconomic aspects. This is true to some extent, but what this shows is that there are some robust laws that apply everywhere."

The researchers analyzed data from about eight million people between 2006 and 2013 in six urban locations: Boston, Singapore, Lisbon and Porto in Portugal, Dakar in Senegal, and Abidjan in Ivory Coast. Previous analyses have used cell-phone data to study individuals' travel paths; this study focused instead on locations and examined how many people were visiting, from how far and how frequently. The researchers found that all the unique choices people make–from dropping kids at school to shopping or commuting–obey this inverse square law when considered in aggregate. "The result is very simple but quite startling," says Geoffrey West, an urban scaling theorist at the Santa Fe Institute and one of the paper's senior authors.

One explanation for this strong statistical pattern is that traveling requires time and energy, and people have limited resources for it. "There is something really very fundamental at play here. Whether you live in Senegal or in Boston, you try to optimize your day," says study lead author Markus Schläpfer of ETH Zurich's Future Cities Laboratory in Singapore. "At the core is the effort that people are willing to invest collectively to travel to certain locations."

Understanding these patterns is important not only for planning the placement of new shopping centers or public transportation but also for modeling disease transmission within cities, says Kathleen Stewart, a geographer and mobility researcher at the University of Maryland who was not involved in the study.

Many researchers estimate travel with "gravity models," which assume that movement between cities is proportional to their population sizes. But these models do not account for travel patterns within cities–information that is particularly critical in tackling disease transmission. Northeastern University epidemiologist Sam Scarpino, who was not involved in the study, says models based on this new finding might better track that flow. For example, New York City residents are more likely to make short, frequent trips within their own borough (such as Manhattan or the Bronx) and fewer trips to a distant borough.

"Those organizational patterns have really profound implications on how COVID will spread," Scarpino says. In a smaller rural location, where many people regularly go to the same church or grocery store, the entire town will experience sharp peaks of infections as the virus sweeps through the community. But in a bigger city, the propagation takes longer, he explains, because mini epidemics can occur in each neighborhood somewhat separately.

Stewart adds: "The authors demonstrate that their visitation law–that takes into account both travel distance and frequency of visits in a way that other models do not–outperforms gravity models when it comes to predicting flows between locations."

About the Author

Viviane Callier is a freelance science writer based in North Bethesda, Md.

Section 3: Urban Inequality

Why Building Green Can Keep People Out of Jail

By Majora Carter

I live in the South Bronx. This small part of New York City receives more than 40 percent of the city's commercial waste. It is home to two sewage plants and four power plants; 60,000 diesel trucks drive through each week. Some 50 percent of the residents live at or below the poverty line. The hospitalization rate for asthma is seven times the national average.

Unfortunately, race and class are reliable indicators of where one can find trees or waste facilities. We see this reality in many ways, from where good public schools and bad ones are found to sentencing disparities for possession of crack versus powder cocaine. Concentrating power plants, truck routes, chemical facilities and waste-processing plants among poor people with less clout results in dirty industrial design. And it happens because the decision makers don't have to live amid their choices.

"Environmental justice" means no community should be saddled with more environmental burdens than any other. Achieving it serves many ends: cleaning up the air, water and soil improves the health and quality of life of local residents and lessens global pollution. If we fix "regional sacrifice zones" such as the South Bronx, we fix a lot of things.

Environmental justice can clean up social problems, too. The U.S. is home to only 5 percent of the world's population, but it houses 25 percent of the world's incarcerated people—in the land of the "free." A Columbia University study linked greater proximity to fossil-fuel exhaust with greater learning disabilities in children—making them better candidates for jail than higher education. Poverty also leads people to do things that land them in jail. A legitimate green economy can provide jobs that reduce these imbalances as well as clean the environment.

I founded Sustainable South Bronx (SSBx) in 2001 to prove that notion. We started by developing a new park where an illegal garbage dump had been. We then created the South Bronx Greenway, redesigning 11 miles of streets into a network that connects neighborhoods and rivers to one another. Our Bronx Environmental Stewardship Training program certifies formerly jobless and often incarcerated people in green-roof installation and maintenance, urban forestry, hazardous waste cleanup and, soon, in retrofitting our aging buildings for energy efficiency. We also work to get laws passed that fuel demand for these jobs. In 2007 we formed a green-roof installation business. And we are now collaborating with the Massachusetts Institute of Technology on the Fabrication Laboratory, a set of computers and fabrication machines that help local designers turn waste into new raw materials.

These and other SSBx programs improve the environment and produce local jobs. The fewer people living in poverty, the less likely they will be victim to the pernicious decision making that got us into the trouble we face today. As these efforts are replicated in other communities, less money will be needed for law enforcement and incarceration, making more money available for education and healthy economic development. Currently many politicians are "talking green," but they are spending more on new prisons than on widespread green economic development.

You can help your elected officials understand the connection between their actions and societal outcomes with four short words: "green jobs, not jails." It's a recipe for success with multiple benefits. Major green trophy projects provide good ribbon-cutting ceremonies, but a well-supported environmentally conscious economic agenda, a Green New Deal, will keep dollars in our local economies, carbon out of the atmosphere and healthy families together.

About the Author

Majora Carter started Sustainable South Bronx in 2001 to clean up her overly polluted neighborhood.

Share the Wealth: New Urban Poverty Atlases Now Provide Data to Slum Dwellers

By Larry Greenemeier

Economic opportunity has always been a big part of the allure of urban life, yet most cities are at least pockmarked by areas of extreme poverty. Often the scope of the problem eludes government agencies as well as the impoverished communities themselves. Poverty atlases that map the extent of privation have existed for decades as a means to alert urban leaders to areas lacking basic services, such as water, electricity and sanitation. More recently, however, groups are looking to deliver this information beyond bureaucrats, going straight to the residents in an effort to empower them to take charge of devising and implementing long-term solutions to their problems.

Since 2007 nonprofit CHF International has been mapping urban poverty in areas of India and Africa through a program called Slum Communities Achieving Livable Environments with Urban Partners (SCALE-UP). "There isn't a lot of granular information on poverty and slums in the cities we're working in," says Brian English, country director of slum upgrading, urbanization and climate change initiatives in India for CHF, which was founded in 1952 as the Foundation for Cooperative Housing to provide affordable homes for low-income families in rural and urban America.

CHF's goal in mapping slums is to provide a more complete picture of why poverty exists in certain areas and how conditions can be improved. For now, SCALE-UP, with the help of about $9 million in funding from the Bill and Melinda Gates Foundation, focuses on three cities in India–Bangalore, Nagpur and Pune–as well as three in Ghana–its capital Accra and its twin port cities Sekondi and Takoradi. Some mapping work has also begun in Haiti.

Data Deficit

There are several reasons why information about slum neighborhoods may be lacking. Often, when a local government designates an area as a slum it has certain obligations to that locale, such as providing access to health care, water and sanitation services. "So until the government is ready to accept their obligations to provide services to these populations they won't collect any details on them or include them in official statistics," English says. "Another way of saying this is policy makers often turn a blind eye to impermanent settlements."

In addition, high-poverty neighborhoods are spreading faster than some cities can track them, particularly if local governments are relying on outdated census information to allocate resources. And government-conducted surveys on poverty in developing countries are often considered crude estimates because the poor's mistrust of authorities skews honest answers, English adds.

Map Making

In general, the first stage in creating a poverty atlas is to identify the communities that meet the criteria for being considered a slum and to plug their locations into geographic information system (GIS) software. (Esri's ArcGIS has been used in Ghana, for example.) This enables CHF and its partners to analyze data and query conditions in a particular community. CHF is likewise using satellite imagery and survey-grade handheld GPS devices to ensure the accuracy of its maps. "We've also experimented with open-source platforms like Walking Papers that allow anyone to add detail to maps and upload them to wiki-style, Web-based OpenStreetMap," English says.

CHF's Pune atlas, created with help from local NGO Maharashtra Social Housing and Action League (MASHAL), is one of the organization's most complete projects, with 477 slum neighborhoods identified. CHF has also been able to drill down and perform socioeconomic surveys within 360 of the slums, according to English. Within those, 85,000 households have responded to a questionnaire, representing roughly 430,000 individuals. To get this more detailed information, CHF broke each slum down into clusters of 25 households and asked a volunteer from each one to gather information concerning the people living in their household. For example, households were asked if they had their own toilet, used a public toilet (and whether it was free) or defecated in the open.

Defining Poverty

Knowing, however, that the wheels of government often turn slowly, CHF also shared the information with the residents themselves. "We gave the volunteers from each household who collected the data summary statistics that they could bring back to their communities," English says. The idea was to enable households to better understand their communal problems and self-organize in an effort to improve their conditions.

In one case in Pune, a group of households sharing information learned that the husbands in their community were drinking heavily and then abusing their wives. They also determined that the majority of the liquor was coming from a single store, which they were later able to shut down to help alleviate the problem, according to English. About 85 projects–constructing drainage pipelines, setting up classes for school dropouts and establishing a library for women, to name a few–have been implemented in Pune as a result of information-sharing efforts.

A distinct characteristic of SCALE-UP is its broad definition of poverty. In Ghana urban poverty has traditionally been a measure almost exclusively of income. CHF's poverty map for Sekondi–Takoradi, the country's third-largest urban area, takes into account access to housing; room and housing density levels; solid-waste services; sanitation and water; and income, says Ishmael Adams, acting country director for CHF in Ghana. The Sekondi–Takoradi study, published in February 2010 with help from the Sekondi–Takoradi Metropolitan Assembly, found that in one of the slum areas studied, among 400 houses there is only one house with a toilet. The rest of the 9,000 residents share a public latrine built in 1958.

The poverty atlas generated significant interest from the mayor of Sekondi–Takoradi, according to Adams, who adds that data in the poverty atlas was used to prepare the city's current Medium Term Development Plan. "Because of CHF's poverty atlas, and the bright light it has shed on several deprived slums, over 30 new projects are now in the works to support the development of the most in-need areas," he says. This includes the construction of water and sanitation facilities at Ngyersia, Kojokrom, Kwesimintsim and New Takoradi—all slums in the Sekondi–Takoradi metropolis.

"This project is not just about Ghana nor is it just about India," Adams says. "This is an approach that is going to have to be taken around the world, and many other similar organizations are trying to do the same."

Plotting Better Cities

Of course, CHF International is not the only organization to create urban poverty atlases. The United Way of New York City and the Community Service Society of New York created a map of that city in 2008. The U.S. Census Bureau has a map of the country based on 2005 census data, whereas the World Bank has a comprehensive series of maps for countries worldwide.

"The power of this project is that the data is not just extracted and put in presentations for policymakers—it's also given back to the communities themselves and used to empower them," English says. "We tell them where to push in their local governments and if there is no place to push, we tell them how to mobilize their own resources to do this." Slums are not the result just of urbanization but also of failed policies and practices of marginalization and exclusion of the poor, he adds, "so it is quite possible to not have slums although it's hard for cities to imagine it."

About the Author

Larry Greenemeier is the associate editor of technology for Scientific American, *covering a variety of tech-related topics, including biotech, computers, military tech, nanotech and robots.*

Street Markets and Shantytowns Forge the World's Urban Future

By Robert Neuwirth

The women maneuvered their crude canoes down narrow alleys of brackish water. They dipped their paddles lightly, gliding slowly past scrap-built houses elevated on spindly sticks that held the structures just beyond the reach of the tide. Here and there a head popped out of one of the homes to check who or what was passing. In the small harbor where the women beached their boats, the shoreline was a work in progress. People were filling the shallows, tamping down layers of trash to reclaim solid ground from the murky brown. Nearby, under a thatched-roof pavilion on one of those pounded patches stolen from the sea, a woman lit a match and put it to a pile of wood chips and sawdust at her feet. A lazy haze of smoke rose into the dusty air.

Greetings from Makoko, one of the most notorious squatter communities in one of the most notorious cities of the world: Lagos, Nigeria—a metropolis caught in a vortex between modernity and misery. With hundreds of ATMs, scores of Internet centers and millions of mobile phones, this bustling, maddening, overjammed city of between eight million and 17 million (depending on where you draw the lines and who does the counting) is fully plugged into the global grid. A hyperentrepreneurial international trading center and the commercial capital of Africa's most populous country, Lagos lures an estimated 600,000 new arrivals every year. Yet most neighborhoods, even some of the very best, have no water, no sewers and no electricity. Makoko—part on land, part hovering above the local lagoon—is one of the mega city's most deprived communities.

Neighborhoods like it—squatter communities—exist all over the world. Rio de Janeiro's 600 favelas dip down to Guanabara Bay and gallop up the steep hillsides from the famous beaches of Copacabana and Ipanema. Favelas are home to 20 percent of the city's residents. Mumbai's countless *jhopadpattis* line the banks of foul-smelling Mahim Creek, sit on the sidewalks of Reay Road and push against the tracks of the city's commuter rail lines. Half of all Mumbaikars are squatters, living on land they do not legally own. Kibera in Kenya–home to perhaps one million, making it one of the largest mud-hut neighborhoods in sub-Saharan Africa–is a short jaunt from downtown Nairobi, but it has no electricity, no sewers and no sanitation, and people who live there pay as much as 20 times what people in legal neighborhoods pay for drinking water.

Although 800 million to 900 million people–one in seven on the planet–live in such places, governments the world over have long looked down at these communities. When they have not been bulldozing or demolishing them, they have acted as if such places do not exist. One example: for decades, the official land-use map of the Nairobi city council showed Kibera, which is perhaps 100 years old and home to as many as one fifth of the city's residents, as a forest, not a neighborhood. With no city services and governments stuck in denial, these places have, of necessity, become hives of inventiveness, industry and self-made enterprise. Despite the hardship and deprivation, such illegal communities are the crucibles of our global future. Governments need to embrace them, not disown them.

Floating Markets

When your neighborhood sprawls over the water like Makoko, you cannot just open your door and walk to the store. Instead the products have to come to you, and the women sliding by on the listless Lagos Lagoon are the waterfront equivalent of a street market. Some carry staples such as *garri* (toasted, fermented cassava), *fufu* (another starch most often made from ground yam), bread and rice. Others sell soda and beer. Still others bring brooms and household supplies across the water.

Their canoes are manufactured by local artisans who sculpt the rough boards by hand to ensure that they can withstand the corrosive seawater. The houses, too, are a cottage industry, built by specialists who know just how far to pound the stilts and just how much weight those flimsy supports can hold. Filling in the shoreline is also an organized operation: just ask the young men who arrive several times a day in boats piled high with silt, which they spread over the compacted garbage.

And that smoky fire–which, if untended, could engulf the entire community–is a business, too. Here operating three large grills without a government license on land that until recently did not exist, Ogun Dairo smokes fish. She does not catch the fish herself but buys them from a cold-storage facility near her home. Nor does she sell the fish herself. She simply tucks the tail into the mouth, creating a compact, ring-shaped item that does not have to be flipped as it is smoked, exposes it to the smoldering chips for a few hours and then packs it in boxes. Those boxes–she typically fills between five and seven a day–go to a distributor who, in turn, sells them to women (the street sellers of smoked fish are always women) who peddle the catch all over the city. "The profit margin is not all that much," she said, using terms that a small businessperson anywhere in the world would understand. "The profit is made by the turnover."

I asked where the fish came from, figuring she would say that the lagoon, so polluted right there, was cleaner upstream or farther out to sea. Or perhaps the fish came from elsewhere in Africa, from other nations up or down the coast, or even from far inland. But I was not prepared for her answer: "Europe." The fish were caught in the North Sea, frozen and shipped to Lagos, where they were transported from the port to one of the most noxious neighborhoods to be smoked and then sold for a few naira of profit each (a few U.S. pennies) on the roadsides of the mega city.

These businesses are not registered, not licensed and not counted in official employment statistics. They lurk in the political and economic shadows. Yet they have become the global norm. Today more than half the workers of the world, or approximately 1.8 billion people, earn their living off the books. And their numbers are growing. By 2020 the informal economy will encompass two thirds of the global workforce, according to the

Organization for Economic Co-operation and Development. What is more, estimates are that almost half the world's economic growth over the next 15 years will come from the top 400 cities in emerging economies. The urban center of gravity–indeed, the global center of gravity–is shifting to the developing world, and these massive do-it-yourself street markets and self-built neighborhoods are a vision of the urban future.

On Their Own Initiative

To planners and government officials, that sounds scary. They worry that ungovernable neighborhoods and off-the-books enterprises will become metastatic, spreading disorder, dysfunction and even outright criminality, dragging entire cities over to the dark side. And as residents themselves will openly admit, living over a polluted estuary or without running water is not the 21st-century ideal. "It is not our wish to be here," Erastus Kioko told me as darkness enveloped his one-room dwelling in Kibera. "If we had money, we would not stay here." He stared at the misshapen mud walls of his home, then added, "I cannot say I have a future."

Still, despite the difficulties, his prospects are actually better in Kibera than if he lived elsewhere in Nairobi. That is because the cheapest single-room apartment in a legal neighborhood of the Kenyan capital generally costs four times more than the average mud hut in Kibera. Sadly, no government or private developer is prepared to build housing that Kioko or almost anyone else in Kibera (or, indeed, in all shantytowns the world over) can afford. Only squatters, building for themselves, have the ingenuity and desire to make these communities work.

In the developed world, people leverage their wealth to get mortgages that enable them to buy materials, hire contractors and build their homes all at once. Squatters do not have that luxury. Their mortgage is the time they are willing to put into building and rebuilding their homes. In Mumbai, hut residents sometimes spend years making and remaking their homes one wall at a time–and scavenged billboards, rusty fence posts, salvaged bricks and half-worn tiles are all valuable resources.

When governments deny these communities the right to exist, people are slow to improve their homes. For instance, when authorities in Rio de Janeiro waged war on the favelas back in the late 1960s, people feared that they would be evicted or burned out of their homes and were slow to invest. Most of the favelas remained primitive—scarcely different from the mud and wood huts of Mumbai and Nairobi. But as the politicians dialed down the hostilities and began engaging with the communities, the favelas rose into the open.

With acceptance, residents rapidly ripped down the old shacks and replaced them with multistory homes made from reinforced concrete and brick. Fly-by-night installers—called *gatos*, or cats—offered favela residents the opportunity to steal electricity from municipal lines (and you can still see their handiwork on utility poles topped with teased-out tresses of wires). Starting in 1997, however, the local power company recognized that squatters do not want the diminished service and short circuits inherent in pirated hookups. Today the utility has struck a bargain with many communities, offering to wire favelas as long as residents accept meters and pay for the electricity used. The program has been a huge success. Having stable electrical service has also worked wonders for public health, because squatters in Rio use plastic pipes and electric pumps to pilfer water from municipal mains. That may also be theft, but it has provided more than a million people in the city with access to safe drinking water.

Under the Umbrellas

Just as squatters have been building the neighborhoods of the future, street vendors and other unlicensed operators are creating the jobs of the future. No government, no global nonprofit, no multinational enterprise can seriously claim to be able to replace the 1.8 billion jobs created by the economic underground. In truth, the best hope for growth in most emerging economies lies in the shadows. In Lagos, for instance, street markets have scaled up into huge roadside enterprises. Alaba International Market, Ikeja Computer Village, Ladipo, and the Auto Spare Parts and Machinery Dealers Association Market have established sophisticated networks for international trade. Merchants voyage to far-flung places (these days most trade is with China) in search of products and profit. They import most of the mobile phones, consumer electronics and car parts sold in the country–and their businesses have burst the boundaries usually associated with street operations. Remi Onyibo and Sunday Eze, two of the leaders of the merchants association in Alaba, told me that the market does more than $3 billion in business every year.

Given that kind of economic power, many major corporations have recognized that they, too, can harness the might of unlicensed entrepreneurship. The mobile phone industry is a good example. In Nigeria the mobile market is led by multinationals such as MTN (based in South Africa), Zain (based in Kuwait) and Globacom (based in Lagos but offering service through much of West Africa). These multibillion-dollar outfits make most of their money by selling phone-recharge cards through a huge, haphazard force of street vendors in impromptu booths under umbrellas at the side of the road. "The umbrella market is a very, very important market now," said Akinwale Goodluck, now corporate services executive for MTN's Nigerian operations. "No serious operator can afford to ignore the umbrella people."

Indeed, one umbrella stand operator told me that there was good money in the trade. She started with just $34 in recharge cards and within six months had increased her business 60-fold, netting a profit of $270 a month, five times the minimum wage then established by the government. But successful and responsible vendor though she is, the multinationals whose cards she sells keep their distance. They sell the cards to distributors who resell them to the umbrella stand operators and claim that this potent street force is actually an army of independent contractors with whom they have no relationship and for whom they have no responsibility. And sadly, the Lagos state government has embarked on a campaign to destroy many roadside markets. This campaign has made it much more difficult for umbrella people and vendors of all sorts to do business on the street. This policy seems punitive and counterproductive given that, according to the government's own estimates, 70 to 80 percent of the working people in the city are part of the informal economy.

Still, the firms of the shadows continue to grow. Mumbai's formerly largest squatter community, Dharavi, has its own links to global trade. Here well-established workshops stitch leather bags and sew shirts for sale throughout the world. In Kenya, although they may not have tapped global markets, Kibera residents own or run many successful small businesses. The community's mud roads are lined with stores, bars, beauty salons, bakeries, tea shops and churches (even houses of worship can function as businesses)—and there are Kibera residents who own successful downtown firms. These outfits are innovative socially as well as economically, and many of the most successful entrepreneurs are women, who traditionally have had no opportunity to establish a degree of independence and power.

Out of the Shadows

The standard consensus regarding the informal sector—whether squatter communities or street markets—is that it is a zone of criminal enterprise and the enemy of civil society. Yet apart from occupying land that is technically not theirs, most residents of squatter communities are law-abiding citizens. Similarly, although they may be avoiding taxes, most people who work off the books are productive members of society.

Indeed, one of the largest misconceptions about the informal economy is that it does not exist in the developed world. In truth, throughout history, squatters have been involved in building many of the world's great cities. Most European capitals were once ringed with large shack settlements. A century and a half ago San Francisco made the leap from a sleepy fishing village to a bustling gold-rush city by legalizing thousands of squatters. And in New York City, the Upper East Side and Upper West Side started life as squatter communities, as did many neighborhoods of Brooklyn. In fact, the last major squatter community in Midtown Manhattan,

Sunken Village on West 62nd Street–in between Central Park and where Lincoln Center now stands–was not pushed out until 1904. In the developed world, the public associates the economic underground with criminal enterprises such as drug dealing–but most of the people earning money off the books are involved in benign businesses, from cash-under-the-table construction jobs to street food vendors to fashion designers who ply their wares online.

As a result of the misconceptions, policy makers have tended to create sharp divisions that lead to simplistic solutions: legal and illegal, productive and unproductive, good and bad. These binary oppositions endanger the livelihoods of more than a billion people and threaten to block an important stage in global development. There has to be a middle ground–one that does not endorse all quasi legality but allows multiple ways for economic markets to function.

Alfonso Morales, a professor of urban planning at the University of Wisconsin–Madison, offers one such proposal. Morales, who put himself through graduate school, in part, by working as a street vendor in Chicago, suggests that municipalities the world over should offer business licenses to street sellers–for a substantial fee–in exchange for not pursuing them for their taxes. In the current reality, unlicensed vendors run a risk every time they hit the streets. For them, a license would mean that the police can no longer harass them, and Morales suggests that they will be willing to pay dearly for that protection. The government would get a clear benefit, too–gaining some revenue from these subrosa street merchants who otherwise provide nothing to the public coffers. It would not be a perfect solution–and certain trades, such as food selling, would still have to comply with health regulations and other rules–but it would mark an important shift to engagement instead of criminalization. Morales says, "We need to go from a purely enforcement mentality to a mentality of 'Let's try to enlarge the pie and increase people's share of it.'"

Martha Chen, a lecturer at Harvard University's John F. Kennedy School of Government who works with informal enterprises through her position as coordinator at Women in Informal Employment: Globalizing and Organizing, puts it this way: "We need to come up with models that allow the street trader to coexist along with retail shops and along with large malls. The informal economy is not the problem. It's part of the solution. Street traders, waste pickers, market women: these people really do contribute to the economy and to their cities. How can we manage our cities in a way that has space for them? What we need to do with the informal economy is to figure out how to help it become more productive, more efficient and more effective."

To be sure, governments do not have a great track record in working with people whose survival depends on hiding their homes and earnings from official oversight. The Indian government, for instance, has a cabinet-level commission dedicated to the informal sector, but that has not prevented local governments from pushing punitive policies against squatters and street hawkers. Still, bottom-up initiatives offer hope. Squatter communities and street markets have developed their own cooperative institutions. In Mumbai women in the *jhopadpattis* and informal marketplaces are creating shared savings schemes and joint insurance plans. In the favelas bunches of families join to create mutiroes–mutual construction societies–that share labor and allow them to build their homes together. In Kibera women form "merry-go-rounds" to pool their money, and the stash goes to one member every week–an infusion of cash that has helped many women grow their businesses and develop financial independence. In Lagos every illegal market has a self-governing association and, often, a market court that handles disputes.

For enlightened governments, each of these homegrown institutions offers an opportunity. The *mutiroes* can evolve into cooperative construction firms that abide by building codes; the merry-go-rounds and savings schemes can morph into cooperative credit unions and microlending societies; and the market associations can move into infrastructure investments and provide their own public services, such as garbage pickup and street cleaning. These things may seem minor, but their cumulative impact would be significant. Even if these cooperative ventures stay small, the more institutionalized and permanent they can become, the easier it will be for them to forge fruitful links with government agencies.

Conversely, only by working with local groups can governments bring a level of inclusive development to the most neglected and maligned parts of the urban world. Through a combination of bottom-up and top-down action, squatters and street marketeers can lead the fastest-growing cities into the future.

About the Author

Robert Neuwirth is author of Stealth of Nations: The Global Rise of the Informal Economy *(Pantheon, 2011) and* Shadow Cities: A Billion Squatters, a New Urban World *(Routledge, 2006).*

Race Colors New Residents' Views of Local Businesses

By Erika Beras

A lot of formerly low-income neighborhoods are becoming gentrified. The people moving in bring money–and racial biases. This contributes to them viewing their new surroundings differently depending on the race of those who traditionally live there.

That's according to a study that examined more than 7,000 reviews of two Brooklyn neighborhoods on Yelp, the online, anyone-can-rate-any-business review site. It's in the *Journal of Consumer Culture*.

Researchers looked at reviews of two areas that have in recent years undergone massive demographic changes–Greenpoint and Bedford-Stuyvesant.

Greenpoint is a historically Polish neighborhood that has long been majority white. BedStuy is a historically black neighborhood that in the last decade saw a 700 percent increase in its white population. Blacks are still the majority, but their population is shrinking.

Both sets of reviews viewed the neighborhoods as "up and coming." But established Greenpoint restaurants were seen as cozy and authentic. While time-honored BedStuy restaurants were described as "gritty and sketchy."

Reviewers in white Greenpoint expressed concerns about preserving the neighborhood's culture as it underwent change. Such concerns were noticeably absent in the BedStuy reviews, which were more likely to talk about the neighborhood using words such as "hood" or "ghetto."

Researchers call such attitudes "discursive redlining," a reference to the overt practice of pricing some people out of areas or services. The racial biases inherent in the reviews can influence popular perception–as well as prospective economic investment.

Photos Tagged as Art Linked to Rising Property Prices

By Christopher Intagliata

Urban gentrification is thought to happen like this: artists move into lower income neighborhoods, looking for cheap gallery and living space. "So they move into these neighborhoods, and they make these neighborhoods really pleasant to live in. They kind of create a nice kind of buzzy scene," said Chanuki Seresinhe, a data scientist at the Warwick Business School, in the U.K. Next come the hip cafes and shops. "But what happens is of course then other people are then attracted to move into these areas. And so the demand for these places increases, and housing prices then go up."

So the story goes. But Seresinhe and her colleagues wanted to see if online data told the same story. So they analyzed four and a half million photos uploaded to Flickr, from 2004 to 2013–each geotagged somewhere in London. Then they tallied how many of each neighborhood's photos were tagged "art." And they found that neighborhoods with a higher proportion of "art"-tagged photos actually did see a higher spike in property prices during the study period. The results are in the journal *Royal Society Open Science.* [Chanuki Illushka Seresinhe et al, Quantifying the link between art and property prices in urban neighbourhoods]

Foursquare and Twitter data can also track gentrification trends, as another recent study showed. But the link here isn't necessarily predictive–maybe neighborhoods with rising prices just attract more photo-tagging art enthusiasts. "That's why I wouldn't say, you know, go out and buy a house in an artsy neighborhood, you're going to be rich in five years' time. Because sometimes there could be other factors involved."

But the connection does suggest your photos and social media posts can be used for more than bragging about brunch. "Sharing this data could be so valuable for academic research. We can actually use this data to do something good in the world." Like giving local organizations and urban planners a better look at how and why cities change. All you've got to do, is add a few tags.

–Christopher Intagliata

[The above text is a transcript of the podcast Science, Quickly.*]*

Who Benefits from Public Green Space?

By Mallory Richards

As cities around the world experience the effects of urbanization and increased density, how and where we provide green space to urban dwellers becomes more important. Proximity to green spaces such as parks and biking trails helps to promote physical and mental health, and should be shared equally amongst urban communities. However, like many public goods, new green space can create higher demand for housing in areas nearby, driving up living costs and displacing low-income groups that have been there for decades.

This phenomenon is known as eco-gentrification, and it leads to a situation where only the relatively wealthy benefit from the designation of new urban green spaces. To explore this phenomenon further, I and professor Katie Black of Kenyon College looked into New York's High Line as a case study. The park was created on an abandoned rail line in the Chelsea neighborhood on the West Side of Manhattan. Known originally as the West Side Elevated Line, the railway opened in 1933 and was used primarily by the National Biscuit Company–better known as Nabisco–whose warehouse is now the home of the popular Chelsea Market, at the tracks' southern end.

The railway was operational until the early 1980s, when the rise of trucking depressed rail use, and the line ultimately fell to disrepair. In 1999, two Chelsea residents formed the Friends of the High Line, a nonprofit group aiming to conserve the structure and repurpose it as a neighborhood amenity, culminating in a proposal to convert the existing structure into an elevated green space. The High Line opened in three sections between 2009 and 2014, and now spans nearly a mile and a half from Gansevoort to 34th Street.

Our paper set out to empirically gauge the impact of the High Line's introduction on residential property values, in the context to the overarching social issue of eco-gentrification. We used a regression analysis in order to understand the varying premiums associated with the introduction of green space on home prices by delineating three "buffer zones" to measure the increase on property value by distance. Through our regression analysis, we found that homes within 0.1 miles of the High Line rose in value 35.3 percent more than homes in the range of 0.1–0.4 miles away. In other words, the home price premium associated with green space proximity decreases as distance from the High Line increases.

Allocating green space without displacing the long-term, low-income residents of an area is an important consideration for policy makers as they seek to balance the positive effects of green space allocation and the negative affects of eco-gentrification. In their book *Just Green Enough: Urban Development and Environmental Gentrification*, researchers Winifred Curran and Trina Hamilton pose their "just green enough" theory, examining how an area can become "too green" and thereby exacerbate the gentrification process. Just green enough theory focuses on social justice and allowing communities to design their own environmental initiatives, with keeping the local population intact as an explicit goal.

It aims to uncouple urban greening projects from the telltale markers of gentrification like opulent waterfronts and luxury real estate development. Quality-of-life improvements generated as a result of urban greening initiatives should be enjoyed by the existing residents of a neighborhood rather than pandering to the entry of wealthier, often whiter, households. As our paper shows, neglecting to make this goal an explicit outcome can lead to excessive gentrification as an unintended consequence.

The stark increase in property values resulting from the High Line confirms the inequitable impact of the walkway's introduction. This reinforces eco-gentrification models in which green space access is only affordable to the wealthy, leaving poor communities bearing the weight of environmental disamenities. Furthermore, lack of green space reinforces the urban heat island effect, causing low income areas to experience disproportionately higher temperatures and worse air quality than their wealthier counterparts.

For these reasons, cities must take a more critical and community-centered approach to greening initiatives. Too often, as was the case with the High Line, private funding and planning lends itself to private interests that do not focus on the public which the good is intended to serve. The High Line is a mark of success for real estate developers and New York's tourism industry, but it has also contributed to rising inequity among New York's residents. The affordability of accessible green space should be a primary issue in the future for urban planners, and the retention of an area's existing residents must be an explicit goal of greening projects in order to avoid the pitfalls of eco-gentrification.

About the Author

Mallory Richards works as a research assistant for mortgage backed securities at S&P Global Ratings in Denver, Colorado.

Section 4: Imagining the Future City

Revolutionary Rail: High-Speed Rail Plan Will Bring Fast Trains to the U.S.

By Stuart F. Brown

America is an absurdly backward country when it comes to passenger trains. As anyone who has visited Europe, Japan or Shanghai knows, trains that travel at nearly 200 miles per hour have become integral to the economies of many countries. With its celebrated Tokaido Shinkansen bullet trains, Central Japan Railway has for the past five decades carried billions of passengers between Tokyo and Osaka in half the time it would take to fly. A new Madrid-to-Barcelona express train runs at an average speed of 150 miles per hour; since its inception two years ago, airline traffic between the two cities has dropped by 40 percent. In contrast, Amtrak's showcase Acela train connecting Boston to Washington, D.C., averages just 70 mph. That figure is so low because many sections of the Acela's tracks cannot safely support high speeds, even though the train itself is capable of sprints above 150 mph. Think of it as a Ferrari sputtering down a rutted country lane.

There has been a recent push to change all this. Earlier this year the Department of Transportation announced the recipients of $8 billion in stimulus funding designed to spread high-speed rail across the U.S. The 2010 federal budget requests an additional $1 billion in rail construction funds in each of the next five years. And in 2008 California voters approved a $9-billion bond measure to initiate an ambitious high-speed rail network that would connect Los Angeles to San Francisco and, eventually, Sacramento and San Diego.

Questions remain, however, about exactly what kind of passenger system will be built. In the decades since the federal government last pursued rail as a viable way to transport passengers—not just freight—train technology has advanced significantly, with advanced high-speed lines spreading through Europe and, more recently, across mainland China.

And what exactly qualifies as "high speed" by the guidelines of the stimulus funding is open to interpretation. Federal authorities, eager to spread the wealth to as many congressional districts as possible, are financing a bevy of incremental improvements to existing lines. In many cases, these projects will only marginally increase passenger rail speeds.

On the other end of the technological spectrum, some efforts aim to bypass wheels-on-rail systems by using magnetic levitation, or maglev technology, in which passenger cars float above a concrete guideway. Momentum for the technology comes in a number of forms. Although maglev trains have been in development for decades, the first (and, thus far, only) commercial system entered service in 2004. For mountainous regions of the U.S., the technology represents the only viable solution to the problem of steep gradients that would otherwise cripple standard rail lines. And perhaps most important, the technology has received a stunning vote of confidence from the world's foremost experts in building and operating commercial high-speed passenger rail lines.

The Maglev Option

The Central Japan Railway (CJR) has by far the world's largest body of experience in operating high-speed trains, having run the sleek wheels-on-rail Shinkansen bullet trains connecting the population centers of Tokyo, Nagoya and Osaka since 1964. Yet the realities of running the bullet system are now spurring CJR's interest in maglev. Every night a marching army of 3,000 railway workers descends on a 12-mile section of Shinkansen track, scrutinizing the rights-of-way, replacing worn components and assuring precision alignment of the rails. The following night they labor on the next 12-mile section of track. The work never ends.

The company must invest all this costly effort because even small imperfections in the tracks can trigger serious vibrations in the speeding trains. These vibrations, in turn, increase wear and tear on the infrastructure. The deterioration of rails, train wheels and the overhead catenary wires supplying electricity to locomotives increases exponentially with the train's running speed. Truly high-speed rail turns out to be murder on the hardware. If the nighttime maintenance work on the Shinkansen line takes longer than expected, its 309-train daily schedule is thrown into chaos.

Hoping to avoid such difficulties, the company plans to construct a high-speed maglev line called the Tokaido Shinkansen Bypass, which it aims to complete by 2025. Although this would not be the world's first commercial maglev line—a 19-mile shot connecting Shanghai's airport with its financial center opened in 2004—at 180 miles, it will be by far the most ambitious. Yoshiyuki Kasai, CJR's chair, told a gathering of transportation officials in Washington, D.C., last June that maglev would be less expensive than traditional high-speed rail in the long run because of less costly upkeep demands over the life cycle of the system. CJR also says maglev promises to reduce trip times because the trains accelerate and slow down much more rapidly than wheel-on-rail trains can.

More significant for the prospects of maglev technology in the U.S., maglev propulsion allows trains to climb much steeper gradients than standard high-speed rail lines can. It is the only way fast trains could pass through much of the western U.S.'s jagged terrain.

The problem for classic technology is traction. Locomotives' steel wheels can maintain only so much adhesion to steel rails before they start to slip, and the train stalls. Common and unpredictable conditions such as rain, snow, ice and even wet leaves place a limit on the steepness of the grade a train can climb or safely descend. Because of this limitation, grades on railways in the U.S. are generally kept below 3 percent, and grade maximums of 2 percent or less are most common.

Maglev lines, in contrast, have no steel-on-steel contact, so traction does not pose the challenge it does on a wheels-on-rails line. Maglev lines can climb a 10 percent grade, which permits planners to select more expeditious routes when laying out new rights-of-way through hilly terrain.

The technology also allows for high-speed transport in areas that would otherwise remain impassable. The Rocky Mountain Rail Authority recently completed an 18-month study of building two intersecting high-speed train lines running along about 400 miles of Colorado's north-south and east-west interstate highways. It concluded that the trains need to be maglev, because some of the grades along the highways reach 7 percent. "You're going through the Rocky Mountains," says Harry Dale, the rail authority's chair. He also notes that because magnetic forces, not physical adhesion, propel and slow the train, Colorado's "snow and ice problem goes away."

Dale believes that the maglev trains built by Transrapid International, a joint venture of the German firms Siemens and ThyssenKrupp, could do the job. Transrapid is the manufacturer of the Shanghai airport system, which has whisked more than 17 million passengers from Shanghai to its airport at peak speeds of 267 mph. Transrapid's maglev trains use conventional electromagnets; the Japanese, on the other hand, have been researching technology that employs superconducting electromagnets not unlike those found inside the Large Hadron Collider. While the superconducting approach provides greater clearance between train and guideway as a precaution against earthquakes, the magnets must be cooled with liquid helium, an expensive and unwieldy proposition.

The Fast Route

Competing proposals for a passenger train line connecting Las Vegas to southern California further demonstrate just how important maglev technology can be. Urban planners have dreamed of linking Las Vegas to Los Angeles with fast trains for decades. "This is an ideal corridor for high-speed trains because you are connecting one of the biggest entertainment districts in America with southern California, one of the largest population centers," says engineer Thomas Bordeaux, senior transportation manager at Parsons Transportation, an engineering firm in Las Vegas. The cities are 270 miles apart–right in the sweet spot between 100 and 500 miles where train travel is more convenient than either driving or flying. And the land between those two cities is little more than sand and scrub, a blank canvas on which to paint the tracks.

Unfortunately, the Los Angeles basin is flanked to the east by the San Bernardino and San Jacinto mountain ranges. Any high-speed line penetrating these natural obstacles would have to scale grades of up to 7 percent, which is only feasible using maglev technology. The California-Nevada Super Speed Train project aims to do just that, connecting Las Vegas with Anaheim, a large city just south of Los Angeles.

The alternative to maglev technology is to avoid the L.A. basin area altogether. The DesertXpress, as the project is called, would build a traditional high-speed rail line that links Las Vegas to Victorville, a high desert outpost more than an hour and a half from downtown Los Angeles (this assumes no traffic, which is an anomaly in L.A.). While it would not require advanced technology, it also would not take passengers anywhere they would want to go.

The DesertXpress will also fail to connect to the planned California high-speed rail system that will link Los Angeles to San Francisco. The California project was one of the two big winners in this year's stimulus fund giveaway, along with an 84-mile route connecting Tampa and Orlando in Florida. When the stimulus money is combined with the $9 billion secured in the 2008 voter referendum, the California project will have in hand more than a quarter of its $40-billion projected total cost. Construction is likely to begin as early as 2011.

Exclusive Access

Regardless of whether maglev or conventional rail-on-wheels technology is used, an inviolable requirement for safe fast-train operation is having special tracks dedicated to the high-speed trains, no exceptions permitted. That is where Amtrak's pokey Acela line, which shares its route with freight and slower passenger trains, was born to fail.

Another necessity is laying out the track so that there are no grade-level crossings, which is where most crashes happen involving trains and road-going vehicles. Time and time again, people try to drive around a closed crossing gate to beat the train, or pedestrians who are unaware that oncoming locomotives project very little sound in front of them notice a train when it is too late to escape. Depending on a route's terrain, lots of overpasses, underpasses and tunnels may be needed to keep the rest of the world out of the exclusive path of the fast trains.

Why has it taken so long for the U.S. to get onboard with technologies that are already ripe? The short answer: passenger trains have not been a federal priority for quite some time. The nation spent decades building interstate highways and airports; investment in tracks suitable for fast trains dwindled to almost nothing. American railroads became almost exclusively low-speed haulers of heavy freight.

But the recent push for green transportation, along with the realization that the nation's highways and airports are already operating past capacity, could bring fast trains into vogue—at least in a few key regions of the country.

About the Author

Contributing editor Stuart F. Brown has been covering transportation topics such as rail, automobiles, trucks, boats, aircraft and spacecraft since 1984. His work has been recognized by the American Association of Engineering Societies, the American Chemical Society and the Institute of Electrical and Electronics Engineers.

Auto Immune: Cities Convert Streets into Pedestrian, Cyclist and Mass Transit Thoroughfares

By John Matson

In New York City, land of the ubiquitous yellow taxi, cabbies and other motorists find themselves with a bit less room to operate these days.

The city closed several blocks of Broadway in 2009 to create a pedestrian plaza around Times Square—a much-publicized experiment that in February became permanent policy, even though it did not improve traffic flow as much as hoped. The Big Apple has also dabbled in shorter-term but larger-scale street closures, barring cars on a stretch of streets leading from the Brooklyn Bridge to Central Park on a series of summer Saturdays in 2008 and 2009. And on June 7, New York City Mayor Michael Bloomberg announced a somewhat less sexy but nonetheless significant change in the city's infrastructure, instituting dedicated bus-only lanes on Manhattan's East Side to speed transit up and down the island.

Each of the measures instituted in the U.S.'s largest city turns over what had been primarily automotive-ruled space to pedestrians, cyclists or mass transit. In doing so the city follows a trend that has caught on in Europe and Latin America—in some cases long ago—but that has been slow to take root in the U.S. If New York and other American cities such as Portland, Ore., prove to be on the vanguard domestically, the changes there could portend a shift in the way urbanites in the U.S. use their streets in the years and decades to come.

"It clearly is a trend," says Lester Brown, president of the nonprofit Earth Policy Institute in Washington, D.C. "I think there are many cities that are ahead on this. Several years ago I was in Stockholm, and already there were many blocks where cars were banned."

Paul Steely White, executive director of the New York City nonprofit advocacy group Transportation Alternatives, says that the popularity of repurposing urban streets around the world gave advocates more leverage to argue for them in the U.S. "By saying that so many disparate cities around the world, from Bogotá to Copenhagen, are heading in this direction, it made it easier to make the case," White says.

Brown points to Enrique Peñalosa, who served as mayor of Bogotá, Colombia, from 1998 to 2001, as a pioneer in the kinds of changes now reaching the U.S. "He's the one who really redefined things when he took office," Brown says. The city's bus rapid transit (BRT) network, TransMilenio, is in many ways more like a subway system, with its own median-protected lanes and station platforms for loading and unloading passengers. Similar BRT systems have sprung up in places such as Mexico City and Ahmedabad, India, giving over lanes of the road exclusively to bus traffic.

"Thinking about why these kinds of reclamation make sense, I always talk about the spatial efficiency of walking, biking and buses," White says. "When you think about what cities are, they are defined by their density–by definition the space between buildings is limited." The car, he adds, is the lowest density mode of transportation; that is, a person traveling by car takes up far more space than someone on a bicycle or on a bus. "In many respects cities are all the same in terms of supply and demand," White says. "There's always more demand for street use than there are streets."

Looking forward, White sees a more malleable future for the way streets are used. With retractable barriers, city planners can create so-called time-flexible streets, which might be open to vehicle traffic during part of the day and pedestrian-only at other times. "You're accommodating peak use–that could be peak deliveries in the morning and peak pedestrian use during lunchtime," he says. "That's something I think you'll see more of, and something we're pushing for."

About the Author

John Matson is a former reporter and editor for Scientific American *who has written extensively about astronomy and physics.*

Can Suburbs Be Designed to Do Away with the Car?

By David Biello

The new kind of suburb wasn't supposed to be so suburban. Packed into 180 hectares, King Farm in Rockville, Md., filled in a patch of lingering farmland just outside Washington, D.C. The village planners left a broad swath of green down the main road, dubbed King Farm Boulevard, that sported along its sides a mix of different types of housing and amenities, such as shops, within walking distance. Down the middle of the boulevard would be the forthcoming train system that would efficiently shuttle new residents to the Washington Metro's Red Line, thereby linking them with the regional public transit system.

As a result of the new design sensibilities, the Congress for the New Urbanism highlighted King Farm in 2008 as an "exciting" development, and the U.S. Environmental Protection Agency cited it as an example of "smart growth." The planned community checked off all the boxes of the "new urbanist" manifesto: a mix of housing types paired with centrally located amenities, designed for pedestrians and cars as well as public transport–oriented.

Instead of embracing that transportation vision, however, the residents of King Farm and the Rockville City Council recently rejected the proposed transit plan—specifically, any light-rail line that would travel down the swath of green explicitly designed to host such a system.

Transit-ready development may mean nothing if local residents are not ready for public transit. And King Farm residents seem prepared to fight the State of Maryland, which bears ultimate responsibility for the decision and still wants to route any transit system through the community. The battle highlights one of the challenges facing so-called new urbanism as it attempts to steer American life away from the car, which has dominated city planning since at least the 1950s.

The new urbanist movement disdained the automobile-centric sprawl, which locked residents into the use of polluting cars for even the most basic trips. The logic of sprawl saw cities eat up a larger and larger share of the surrounding real estate, fueling habitat destruction, smog from tailpipe emissions, runoff from impermeable pavement and other environmental ills.

Communities inspired by a return to the walkable cities of most of human history have sprung up from coast to coast, including places where one might least expect it, such as Los Angeles, Denver and even Salt Lake City, the latter of which has shifted growth patterns away from sprawl. But, as the example of King Farm shows, new urbanism can devolve into the kind of central planning criticized by the very founders of the urban planning school of thought—and can also result in what is essentially a style without substance. "It is more about lifestyle than ecologically sound cities," says sociologist Saskia Sassen of Columbia University, a member of the U.S. National Academy of Sciences panel on cities.

Planner and developer Jonathan Rose of Jonathan Rose Companies of New York City echoes that sentiment. Not owning a car reduces consumption, he notes, but "that reduction in consumption has to come with an increase in happiness. If it comes with suffering—it takes three buses to get where you want to go—then it's not sustainable."

Such suffering seems to be the case for residents of King Farm, some of whom would choose the freedom to be stuck in Beltway traffic rather than face the peril of a community divided by train tracks. "We've come a long way from an era where many cities produced affordable housing through what gets endearingly called 'drive 'til you qualify' strategies," says Uwe Brandes, vice president of initiatives at the Urban Land Institute, a nonprofit research and education think tank focused on land use. "The area of caution associated with transit-oriented development is there's a danger that these areas in cities can become economic and social enclaves. That's a yellow, blinking light."

About the Author

David Biello is a contributing editor at Scientific American.

Can a Sustainable City Rise in the Middle Eastern Desert?

By David Biello

Oil money has conjured up a pricey experiment in sustainability in a patch of desert between downtown Abu Dhabi and its airport. There, the city of Masdar ("the source") is rising. It is meant to signal a shift away from fossil fuels by hosting a variety of ecofriendly approaches, such as a system of subterranean electric cars—Personalized Rapid Transit—that whisk visitors and residents from point to point. Yet despite its ambitious goals, innovative planning and best intentions, Masdar may likely be only a mirage for other cities hoping to mimic its approaches to sustainability.

The brainchild of ruling Sheikh Khalifa bin Zayed Al Nahyan as funded by Mubadala Development Co. and envisioned by architect Norman Foster, nascent Masdar is certainly a city to behold. The sun's power is harvested via concentrating mirrors or the photovoltaic panels in a 10-megawatt array just outside the city walls. Wells may be drilled to tap Earth's heat, and the entire city will be built in a perfect square raised some seven meters into the air to help capture desert breezes. A 43-meter-tall wind tower that looks like a grated oil-rig funnels these breezes to street level, and those avenues are shaded by their orientation to the sun. The buildings range from undulating red sandstone curves with an Arabic flair that are student apartments to a squat armadillo shell that shelters Masdar's anchor tenant—the Masdar Institute of Science and Technology (M.I.S.T.), a joint venture of higher learning between M.I.T. and the Abu Dhabi government.

M.I.S.T. is now graduating its first crop of scholarship students specializing in renewable energy. For example, in conjunction with aircraft-maker Boeing and Des Plaines, Ill.–based refiner UOP, LLC, M.I.S.T. researchers are studying the potential for salt-tolerant plants to serve as an energy crop for biofuels for jets—and Masdar hosts one of the first sewage sludge–to-power demonstration projects in the world. It also is the future home of the International Renewable Energy Agency.

"Masdar should receive a lot of credit for being explicit about the integration of energy into the planning of urban environments," says Uwe Brandes, vice president of initiatives at the Urban Land Institute, a nonprofit research and education think tank.

Sustainable Elements

But is Masdar truly sustainable? That word has economic, social and environmental implications: the city must help its inhabitants generate wealth; it must make them, at least in their view, healthier and happier; and it must be able to continue to secure the resources–natural and otherwise–to continue. Cities can and do collapse–whether it be the ur-city, Ur, in Mesopotamia or the decline and fall of Rome from a million-inhabitant imperial capital two millennia ago to a village of shepherd's flocks and a miserable 10,000 peasants over the span of 10 centuries (although it has rebounded to be a global metropolis as modern Italy's capital).

"When we talk about sustainability, we're really talking about balance," ULI's Brandes notes. "We're not talking about an end state but a process."

The modern world highlights a slightly different focus for the term sustainable, which the United Nations Brundtland Commission defined in 1987 as "development that meets the needs of the present without compromising the ability of future generations to meet their own needs." That is further elaborated for cities by a set of 10 guidelines–known as the Melbourne Principles adopted in 2002 and ranging from long-term vision to enabling good governance–to determine a city's relative sustainability.

Sterility Guaranteed

Achieving that sustainability requires planning, and the planned city has long been the dream of architects, engineers and others, from the re-creation of Paris by Baron Hausmann in the 19th century to redesigning the city around "green" in more recent decades. However, much like the sterile planned cities imagined by imperial rulers–from the caliph Al Mansur founding Baghdad in A.D. 762 to Brasilia born of an idea from an imperial-era planner and finally built to open up Brazil's interior to settlement in 1956–these new designs aimed at improving the ecological soundness of cities suffer from an inability to be organic. The "garden cities" of the U.K., like Letchworth built outside of London in 1902, quickly faded into the surrounding sprawl as people repurposed the planned communities.

"No matter how creative we are when we design a new city from scratch, it has a sterility to it," says New York City–based urban planner and developer Jonathan Rose. That is certainly true for Masdar, despite the organic shapes of its main architecture. Masdar at present is essentially a walled city for rich students in the midst of an arid, dusty plain–perhaps not a sustainable vision for the bulk of future urbanites. "It really feels like I'm living in a spaceship in the middle of the desert," current M.I.S.T. graduate student Laura Stupin wrote in her blog.

Even worse, Masdar is located in an area that demands it must consume a lot of energy, perhaps to the point where it cannot ever be considered green. "This is being built in a location that in principal is hostile to human settlement," argues Konrad Otto-Zimmermann, secretary general of the International Council for Local Environmental Initiatives (ICLEI), a local government group. "The fact alone that you would build a city where you have to apply and use huge amounts of energy to work against nature–whether that's the way to go I question."

In fact, on a per capita basis, the citizens of the United Arab Emirates are among the largest emitters of greenhouse gases on the planet, thanks to controlling some 8 percent of the world's proved reserves of oil coupled with a desperate need for air conditioning in the brutal heat of the desert. Masdar's neighbors include an airport, a Formula One racetrack and, a bit farther away, an oil refinery–all major sources of greenhouse gas pollution, among other environmental ill effects.

And like most eco-cities—whether Masdar, Dongtan or even Curitiba in Brazil—Masdar's roots lay in fiat, decreed by authoritarian regimes ordering the cities to go green. Masdar has a "green policeman" who shames those for using too much energy as well as shuts off showers after a few minutes. Whether this approach makes sense for other cities and for long-term sustainability remains to be seen.

But Masdar may yet prove a source for planning lessons that can then be applied elsewhere. In fact, if cities are going to save us from future-past environmental perils, it will have to start in the cities that already exist. "It is more a laboratory than an option for the millions of cities in the world," notes sociologist Saskia Sassen of Columbia University. "We need laboratories, but they are by no means enough to address the problem."

About the Author

David Biello is a contributing editor at Scientific American.

The Inconvenient Truth about Smart Cities

By Kendra L. Smith

This week, we learned that tech guru and mega philanthropist Bill Gates purchased 25,000 acres of land in Arizona with the intent to build a smart city from the ground up. The community, called Belmont, will "create a forward-thinking community with a communication and infrastructure spine that embraces cutting-edge technology, designed around high-speed digital networks, data centers, new manufacturing technologies and distribution models, autonomous vehicles and autonomous logistics hubs," according to a spokesperson for Gates' real estate firm Belmont Partners.

This announcement comes on the heels of the release of 2017 Smart Cities Index by EasyPark Group, which highlights cities at the forefront of smart urban growth in such areas as advanced transportation and mobility, Internet connectivity, environmental protection, and several other elements. Given the global investment and public attention devoted to smart cities in recent years, critical explorations of their potential and challenges are worth discussion.

Somewhat of a catchall term for wired, ubiquitous, connected or networked cities, a smart city refers to the technological and data-driven urban systems designed for efficient, resilient and economic growth. They are supposed to be cities of the future that leverage technology and data to improve the lives of citizens and to become more proactive and responsive to the needs of the city. These goals are achieved, the theory goes, by running the city on an integrated operating system where ubiquitous broadband service and sensors master myriad city functions in real time.

The movement toward smart cities signifies attempts by public and private industry to tackle the complexity and, oftentimes, incoherence of urban design efforts to create livable and sustainable communities. However, the potential of smart cities to deliver on their goals should not be mistaken with the reality of smart cities.

The current reality of smart cities is that there aren't any. At the end of the day, most so-called smart cities are just cities with a few or several standout smart projects. Such projects can take shape in a variety of ways. In Pittsburgh, the Pennsylvania Department of Transportation is piloting a $30 million smart-signal system that utilizes adaptive traffic signals to read traffic conditions and make adjustments to keep traffic flowing. Kansas City invested nearly $15 million in a smart lighting project that will install 200 lights along its new streetcar line. The lights have built-in sensors and cameras that detect the presence of people and can turn off when no one is around to save 20–30 percent in electricity costs. Both efforts represent the potential of smart city technology, but they certainly do not represent the networked, end-to-end planning of an entire smart city.

A big reason for the disconnect between smart city potential and reality is the fact that smart cities are where the digital world blends, but can also collide, with the non-digital world. Non-digital issues such as legacy governance, social justice, politics, ideology, privacy and financial elements that are not so smart, efficient or resilient when smart-city planning starts can become large factors. Any one of these elements can pose a challenge in and of itself and grow to monstrous proportions when combined with other longstanding problems in a city. Imagine the entanglements that existing public and private industries must go through to implement a single smart city project, let alone numerous projects such as smart lighting, smart transportation, smart buildings and the like to actually make a more complete smart city. Bill Gates' effort is notable because Belmont is a blank slate to be built from the ground up.

For stakeholders not fortunate enough to build a smart city to their specifications from its inception, finding the best way to implement smart city projects can present a major challenge. The cities highlighted in the 2017 Smart Cities Index are at the forefront of truly making their communities smart but even these cities must contend with the challenges of their past and present. Considering the trajectory of smart city, a few critical questions should come to the fore of any successful implementation effort:

1. How will a smart city affect social justice in my city? Dialogue around smart cities often favors a narrative that they will be places that allow us to thrive by improving our quality of life. Without discussions of how to implement projects in inclusive ways, the efforts can reinforce institutional privileges and protections as well as exclusionary practices. Smart cities will not be utopian societies that, by the sheer presence of their technology, make everything better. As such, we must spend more time on our pressing socioeconomic issues of today to prepare ourselves to create smart and just cities.
2. What does it really cost to develop a smart city? Columbus, Ohio won a $40 million grant from the U.S. Department of Transportation's Smart City Challenge in 2016. By 2017, the city had developed several public-private partnerships that turned the $40 million into $500 million in private funding aimed at supporting a smart city project focused on transportation systems. It is unclear if other cities could pull off such a financial feat. Without government funding and strong partners off the bat, city leaders' ability to raise sufficient funds for a smart city is an uphill battle that cannot be ignored.
3. Who decides what the city really needs and will operate going forward? With a smart city comes a significant amount of decision making on what to do, who will do it, why and when to do it. The answers to the questions are not easy and can have massive repercussions. Take, for instance, the challenge of gentrification and urban displacement, which has long been framed simply as a symptom of wealthier people moving in to communities and effectively nudging out lower-income individuals. However, public investment can play a critical role in this process too. Perhaps the most

shining, unfortunate example of this is what San Francisco Federal Reserve researchers refer to as "transit-induced gentrification" in which public investment in transit–light rail, buses, subway–attracts affluent individuals. So much so that several studies have found that transit investments can alter the demographic composition of the surrounding neighborhood, resulting in pushing out lower-income individuals and creating new problems within the city. Potential outcomes like these should prompt questions about who should be making these decisions about public investments associated with smart cities. Finding pathways to figure out what the public wants from its city (and perhaps more importantly, what it does not) is critical. This requires citizen participation early in the process and throughout. The New Delhi-based Housing and Land Rights Network released a report, "India's Smart Cities Mission: Smart for Whom? Cities for Whom?" The report highlights the massive problems with Indian Prime Minister Narendra Modi's pledge to build 100 smart cities by 2020. Among the problems is the focus on technology of the future instead of issues of the present such as an agrarian crisis, insufficient civil rights for women, forced evictions to make room for the implementation of smart city projects, and so on.

The promise of smart cities is enormous but they require much work to prepare long before network switches are flipped. Creating truly smart cities will require that communities find a satisfactory balance between the present and the future.

The views expressed are those of the author(s) and are not necessarily those of Scientific American.

About the Author

Kendra L. Smith is the associate director of community engagement in the Center for Population Health Sciences at Stanford University. She recently co-authored "Big Data and Smart Cities," a report published by the American Planning Association.

What's Holding Smart Cities Back?

By Amit Nisenbaum

With the global population set to reach 9.7 billion by 2050, two thirds of which will be living in cities, according to United Nations projections, governments worldwide are already facing an array of challenges in ensuring the sustainability, security and success of an ever-more-crowded planet.

The urgency of these challenges underscores the importance of smart cities continuing to fuel growth, innovation and cultural dynamism amid this wave of continued urbanization. To do so, they'll need to harness data analytics, smart infrastructure, sensing technologies and innovation networks to improve civic services, manage day-to-day operations and make cities attractive destinations for capital.

But while the case for smart cities is straightforward, the path to building them is not. Key hurdles stand in the way of unleashing the full potential of smart cities. Here's a look at five of the most pressing challenges that will need to be addressed.

Privacy Fears

Amid fierce debates about how tech giants have made use of data, smart cities have become yet another flash point in the war over digital privacy. Even among those who support smart cities, there are fears that in practice, smart city technology will give Big Brother–be it big government or big tech–too much visibility into citizens' lives.

In Toronto, which has partnered with Alphabet's Sidewalk Labs to build an innovation hub, a privacy expert consulting on the project recently resigned over alleged shortcomings in data privacy protections, sparking calls for more measures against third-party exploitation of individual information.

As cities grapple with urbanization, digitization, and the growing abundance of data, public debate over how to realize the benefits of smart cities without sacrificing privacy will intensify—and while privacy safeguards will always remain a matter of contention, with no neat and easy solution, serious discussion on privacy today will help bring to fruition the smart cities of tomorrow.

Incomplete Data

Even as privacy advocates voice concerns about too much data in the public sphere, there's also a data shortage that demands attention.

For example, all but a fraction of the data used to power connected vehicles are visual data—but to more intelligently control traffic and ensure road safety in smart cities, planners should turn their attention to harnessing data that are not visual. For example, we as humans can "feel" and respond to certain road conditions based on that feel, conditions such as black ice, potholes, windy conditions and so on. Yet autonomous vehicles and the mapping technology required to train them are often missing this element. Compounding the issue further, many methods of visual data collection can be extremely costly, difficult to scale, and inaccurate, as is often the case with data collected from cellphones.

Faster Connections in a 5G World

Boasting fast connection speeds with ultralow latency, 5G will drive smart city development. For residents of Houston, Indianapolis, Los Angeles and Sacramento, commercial 5G service is already a reality, and more municipalities are preparing to enter the 5G era themselves. Although the introduction of 5G will pave the way for a number of benefits, including better and faster communication between municipal systems and services, implementation will be no walk in the park and the difficulties are embodied by the providers such as AT&T coming under fire for "fake" 5G service.

Microcells, the small-cell antennas that enable coverage, will need to be linked up with cell towers that relay signals to end-users once 5G is rolled out. This may entail the construction of millions of new cell towers, which is certain to provoke fierce debates over cost, aesthetics and even public health. To return to our earlier case study, connected cars require high bandwidth to support the collection of visual data–and with millions more of these increasingly sophisticated computers on wheels set to be deployed, cities can't afford to delay addressing the infrastructural hurdles.

Funding

Given fiscal constraints, smart cities can't be bankrolled by governments alone. Public-private partnerships, however, provide a pathway for funding smart-city initiatives and enabling local innovation.

Cisco, to take one example, announced in 2017 a $1 billion City Infrastructure Financing Acceleration Program, in partnership with a group of financial firms, to jump-start technological innovation in smart cities. As population growth strains infrastructure and places greater demands on local authorities, we'll need more of such partnerships to play a vital role in supporting smart cities' growth. These public-private initiatives can also help overcome public resistance to the costs of major investments like high-speed rail.

Data Storage

Smart cities will run on big data, and those data will need to be stored: no cheap feat.

IT storage spending worldwide is projected to reach $51.3 billion in 2019, and ensuring the security of municipal data–a paramount priority, particularly given privacy concerns–will require additional investments and protocols, while new investments in edge computing will enable data to be processed more efficiently.

All this data storage comes with an additional cost: to the electric grid, and more broadly, to the environment, which absorbs increasing carbon dioxide emissions due to the massive amounts of energy data centers consume. While more sustainable mobility and energy ecosystems are key benefits of adopting smart-city initiatives, these benefits can be magnified with the use of green solutions that are easier on the grid.

Forging a future in which our cities are smarter, safer, greener, more competitive and more livable will require public and private players to come together to tackle the technological, economic, environmental and political costs associated with smart-city innovation. The return on these investments of time and resources will be tremendous. For the sake of the globe's long-term health and well-being, there's no time like the present to get to work.

The views expressed are those of the author(s) and are not necessarily those of Scientific American.

About the Author

Amit Nisenbaum is CEO of Tactile Mobility.

Section 5: Designing the Green City

The Fishy Business of Waste

By Madhusree Mukerjee

Fish love sewage, and the inhabitants of Calcutta love fish. These city dwellers also produce sewage: 700 million liters of it a day. By growing fish on this waste, fisherfolk to the east of the Indian metropolis discovered how to link these needs at the turn of the century. But this natural recycling system–the largest of its kind and possibly the oldest–is threatened by a state government that sees profitable real estate in the expanse of shallow waters.

The fisheries flourish on elaborate folk technology. Diverted into a series of ponds, the sewage sheds its putrescent solids. Air-breathing fish and water plants survive in pools nearest the sewage inlet. Further downstream, algae bloom on nutrients in the clear water. Tilapia and carp, in turn, thrive on the algae. More than 20 tons of these fish feed the Calcutta market every day. (Bioassays of these fish show that their levels of heavy metals and coliform bacteria are low.) The algae also provide much needed oxygen to the heavily polluted city.

Further, sludge from the distillation ponds fertilizes extensive "garbage gardens." Calcutta's solid waste, stripped of its paper, plastics and metals by some 25,000 ragpickers who roam the city's noxious dumps, is ultimately almost entirely organic. Irrigated by treated sewage and then composted, the waste becomes soil on which farmers grow a variety of vegetables. Their harvest supplies the city with 150 tons of produce a day.

This recycling expanse, however, has been shrinking as Calcutta expands eastward. In the 1960s northeastern water bodies were filled in to create a township. The Dutch consultants who drew up the plans apparently overlooked what British engineers had noted in the 1880s: the land naturally slopes to the east. Once these pond beds were gone, the eastbound sewage canals stopped flowing during the rains of the monsoon season, and now they regularly flood the city.

In addition, thugs–or antisocials, as Indians call them–associated with the communist state government are reported to have seized many fisheries on the pretext of distributing land to impoverished laborers. Instead the parcels have been sold to various real-estate developers.

Recently citizens groups sued to stop the government from building a shiny new World Trade Center, funded by a multinational group, on such "reclaimed" terrain. The case is still being disputed. Meanwhile politicians are touting a "Megacity 2011" expansion project, the logistics of which, observes Asish K. Ghosh of the Zoological Survey of India, have not been divulged to environmentalists.

One thing is clear, though: the plan involves grand designs on the fisheries. At the same time, these water bodies are becoming ever more essential. In February the Indian Supreme Court threatened to close 30 factories that were discharging pollutants into the Hoogly River, west of Calcutta. Dhrubajyoti Ghosh of the Institute for Wetlands Management has suggested that these effluents could be treated by expanding the fisheries.

But the government is clearly not listening. And Calcutta's leap into the 21st century could well become a plunge into miasmic waters.

About the Author

Madhusree Mukerjee is Scientific American*'s senior editor for science and society.*

How Green Is My City

By David Biello

It was to be the ultimate urban paradise. Hundreds of pages of plans, maps and charts detailed the construction of a state-of-the-art eco-city called Dongtan on China's Chongming Island, at the mouth of the Yangtze River. Energy-efficient buildings would be clustered together to encourage residents to travel on foot; only battery- or hydrogen-powered cars would be permitted in the development. Surrounding organic farms would supply food; sea breezes and the burning of husks of China's staple crop, rice, would furnish power. Canals and ponds would incorporate the local wetlands, providing restful views for humans and continued respite for migrating birds.

Yet for all its grand goals, this island city-to-be remains unbuilt. Whether China has abandoned the project totally is unclear. It was originally slated for completion in 2010 but has failed to proceed beyond the construction in 2009 of a tunnel and bridge linking Chongming to the mainland. It is one of numerous planned eco-cities around the world that have fizzled, many because of cost. Even if every planned eco-city were successful, however, their effect on overall energy use and emissions would be minimal because the vast majority of urbanites would still live in existing cities. All these reasons suggest that we cannot rely on new construction to fully address the challenges of feeding, housing and transporting urban populations in ecologically sound ways. We need another solution.

The solution needs to take the future into account. Today's cities are by many measures greener than suburbs–among other things, urbanites use less energy and emit less carbon dioxide per household than their suburban counterparts do because they live in closer quarters and use public transportation. But it is not enough to be green. Cities need to be sustainable, too. That is, they must be able, as the United Nations's World Commission on Environment and Development stated in 1987, to meet "the needs of the present without compromising the ability of future generations to meet their own needs." Existing metropolises will not be able to sustain themselves if left to operate on a business-as-usual basis–demand for resources will outstrip supply as the number of people inhabiting cities swells from more than three billion today to more than six billion by 2050. The many traditional cities that are mushrooming in China and India and elsewhere are facing the same conundrum.

In theory, new cities could have sustainability built into their infrastructure from the start–as was planned for Dongtan. But a larger payoff would come from retrofitting existing cities for sustainability, given how many there already are. "We must work with existing cities; I have no doubt about it," asserts sociologist Saskia Sassen of Columbia University, who has spent her career studying cities. That approach would be less costly than rebuilding cities from scratch and could still conceivably save enormous amounts of energy and water, allowing today's cities to flourish for centuries to come. To meet these objectives, engineers, city planners and locals could take good ideas from planned eco-cities that have failed as cities but succeeded as incubators for innovation. Simple changes, such as training building superintendents in best practices, can also go a long way toward helping cities support us well into the future.

Saving Energy

A key priority for cities adapting to a world transformed by global warming is increasing energy efficiency and reducing greenhouse gas emissions to stave off even more catastrophic climate change. "As the primary centers of economic activity globally, cities are significant consumers of energy and emit nearly three quarters of the world's carbon emissions," New York City mayor Michael Bloomberg told a recent conclave of mayors at a meeting of C40, a planning group for 59 major cities engaged in efforts to combat climate change.

A major focus of C40 is equipping old buildings with energy-efficient features. In the U.S., the average building–whether skyscraper, house or church–was built in the 1970s. Replacing their black-tar roofs with white roofs that reflect sunlight to keep buildings cooler in the summer or installing solar-thermal hot-water heaters, for example, can translate into major energy savings: heating hot water accounts for 17 percent of the energy used by buildings in the U.S., according to the Department of Energy. C40 has thus partnered with the World Bank to ensure funding for such retrofitting projects, among other climate action plans for cities.

Existing cities might also benefit from installing transportation systems originally conceived of for planned eco-cities. Tailpipes in the U.S. spew 1.7 billion metric tons of carbon dioxide a year, along with a host of noxious fumes. In contrast, the electric car system proposed for Fujisawa City in Japan would produce no tailpipe emissions. Electric car systems require infrastructure, though, particularly to ensure that people can charge the cars. In Tokyo a company called Better Place has had success in testing a system of electric vehicles powered by batteries that, when depleted, can be quickly and easily swapped out for recharged ones at battery switch stations. In the near term, simple changes, such as converting buses to run on compressed natural gas rather than diesel, can both clean up the air and improve efficiency. Already such efforts have helped Denver save more than 24 million gallons of gasoline between 2005 and 2009.

Cities must not only conserve energy and limit emissions but also diversify their energy supply. New York City recently mandated a switch from heavy heating oils to lighter, cleaner-burning fuels, such as natural gas, in a bid to improve air quality. Yet even such seemingly straightforward decisions can demand difficult trade-offs: David Bragdon, director of Bloomberg's Office of Long-Term Planning and Sustainability, notes that New York is struggling to reconcile this increase in the use of natural gas with its desire to prevent hydraulic fracturing, or fracking–a process for producing natural gas from deep rock–in its watershed because fracking can contaminate water supplies.

Water and Waste

Ensuring that sustainable supplies of freshwater continue flowing to growing urban populations is another daunting task facing the international community. Large swathes of the world are already pushing the limits of water availability. Cities throughout the western U.S., from Denver to Phoenix, for instance, are using up more than the normal flow of the Colorado River. And the International Food Policy Research Institute estimates that about half of global grain production will be at risk because of limited water by 2050. To help cities conserve, C40 has developed a list of best practices based on case studies of strategies employed by cities ranging from Austin, Tex., to Tokyo. Austin, which launched its water-efficiency program in 1983 in response to a housing and commercial boom, offers a number of incentives to curb water use, including rebates for installing rainwater-harvesting systems and water-conserving toilets. Tokyo, meanwhile, is the world leader in detecting and controlling leaks in its waterworks. It has earned this distinction by systematically checking, repairing and replacing pipes and by fixing leaks on the same day that they are identified.

For its part, the planned city of Masdar in the United Arab Emirates (not a C40 city) takes a Big Brother–like approach to conserving water: showers shut off automatically after a few minutes, and each resident's water use, along with energy use, is monitored via a computerized smart grid that allows the provider to intervene if users get greedy.

Water must also be clean. For most cities, meeting this objective will mean not maintaining the status quo but vastly improving on it: according to the U.N., nearly a third of city dwellers live in slums, which typically lack access to safe drinking water and sanitation services, leaving them vulnerable to cholera and other waterborne diseases.

Poor waste management is not just a problem for water quality, however. New York City, for example, has closed its landfills in Brooklyn and Staten Island and now pays as much as $100 a ton to move waste hundreds of kilometers away. Even recycling is not a panacea—Dubuque, Iowa, halted its glass recycling program, according to Mayor Roy Buol, because trucking the material to far-flung processing plants added more to the city's greenhouse gas emissions than dumping it in a landfill. Even better than simply disposing of waste or recycling it, of course, would be making something useful from it. Just such a transformation is taking place in an industrial park outside the city of Rizhao in China, where Luxin Jinhe Biochemical Company makes citric acid for beverages from cassava, corn and sweet potatoes. The leftover waste flows into tanks called biodigesters, where microbes turn it into solids that can be converted into meal for animal feed and methane that can be burned for industrial purposes, such as generating electricity. In fact, capturing methane from landfills is one of the cheapest ways to cut down on greenhouse gas emissions while making a new "natural" resource.

Easy Fixes

Without a doubt, existing cities will need cutting-edge technology to help achieve their long-term sustainability goals. But policy tweaks and low-tech solutions can play an important role, too–for instance, changing building codes to require more energy efficiency, which could be achieved with better insulation. Indeed, the real battle to make an existing city such as New York more sustainable may be won in the minds of superintendents managing the metropolis's roughly one million buildings. Hence, the U.S. Department of Energy's Green Supers program, which trains building service workers in green operations and which recently graduated its first class. "I was under the impression that these techniques were very expensive. It's just time, it's just dedication and just applying it," said superintendent Victor Nazario during his address to his fellow classmates at commencement.

These concepts are spreading worldwide, thanks to organizations that bring leading cities together to share plans that work, such as C40 and ICLEI–Local Governments for Sustainability. And when cities act, national governments notice–taking its cue from the 259 cities in China that are striving to be low-carbon, the Chinese Ministry of Housing and Urban-Rural Development is now studying plans to encourage the use of more energy-efficient and long-lasting building materials, which could significantly enhance the sustainability of the country's boomtowns.

Cities are an expression of our collective will, a potent mix of economics and environment, private visions and public dreams. Boosting their ability to provide clean energy, transportation, food, water and waste disposal will be key strategies to ensuring a brighter future for humankind. But when it comes to eco-cities, those efforts too often prioritize aesthetics over the real-world needs of people. And it is the people who ultimately make a city sustainable or not.

About the Author

David Biello is an associate editor at Scientific American *Online.*

How Businesses Use Data to Outsmart Nature's Wrath

By Lou Gritzo

Floods, typhoons and earthquakes are part of life on this planet. Yet it's frequently a shock when *your* community, business or nation is struck. More often than not, the destructive aftermath of a disaster is attributed to fate or inevitability.

But the majority of these losses are actually preventable. You could have better fortified your property. If you had the option, you might have selected a less risky location. And if it was your business that was affected, you could have more deeply explored the risk you were inheriting from all of your locations, suppliers and customers.

This thinking is at the center of my company's ongoing effort to quantify natural hazard risk around the world. As a commercial property insurer, we do this to help businesses decide where to site new offices, warehouses, or plants or for companies to assess and prioritize their existing risk profile. It's important for businesses to be resilient—physically, economically and operationally—so they can stay in business and succeed relative to their competition, and supply economic well-being to their communities.

Fortunately, there is an expanding world full of data that we've analyzed to help these companies, by ranking countries according to their natural hazard risk.

Here's how the analysis came together and what it showed.

Natural Hazard Exposure Population Centers

First, we use our global hazard maps, including our newly released global flood map to assess the fraction of each country exposed to natural hazards. Since certain areas may be exposed to natural hazards that are not inhabited, analysts next determine how much of a country or region's population is in harm's way. They examine satellite photographs taken at night to identify a country or region's most populous areas by the densities of their lights. Population signifies economic activity. Analysts then calculate the percentage of the country's area devoted to economic activities that is exposed to at least one natural hazard peril (earthquake, wind, or coastal or riverine flood).

Furthermore, in their day-to-day work, our 1,800 property engineers perform more than 100,000 location-based risk assessments annually around the world. Part of the assessment includes determining the extent and type of natural hazard exposures that are present. We combine this data with the fractions of populated areas exposed to hazards to produce a refined ranking of exposure to natural hazards. This ranking is available here.

This analysis forms the basis of natural hazard exposure rankings we've developed for 130 countries and territories. Exposed areas are determined based on potential losses from wind gusts greater than 100 mph (161 kph), water flowing from rivers in 100-year flood zones, or more-frequent-than-500-year earthquake motions that can cause damage to weak systems.

Two countries, the United States and China, are each divided into three regions to account for different dominant natural hazard exposures within these countries. In the United States, for example, the leading natural hazard on the West Coast is earthquake, and on the East Coast it's wind.

Next we ask the question: What can make a natural hazard worse, all things being equal?

Urbanization Rate

Urbanization, though associated with economic growth and opportunity, can be risky. When urbanization is rapid and unplanned, it poses acute risks to a country's critical infrastructure and social stability. Rapid urbanization can mean high stress on water supplies, power grids and other infrastructure.

In such instances, many cities are expanding faster than core infrastructure, utilities and drainage systems can be planned, developed and managed. This stress would make the infrastructure a weak spot if the region were struck by natural disasters such as windstorms, flood and earthquakes. Flood events, especially, are prone to be more severe since extensive urbanization hardens landscapes, potentially increasing flood runoffs.

Urbanization rate is defined as the average annual rate of change in the extent to which a country's population is living in an urban area. Our ranking of countries based on urbanization rate is available here. Our source for this data is the United Nations.

(Since urbanization rate can be an economic risk as much as a property risk to a business, we consider it separately from natural hazard exposure. Both drivers–urbanization rate and natural hazard exposure–contribute along with 10 other drivers to an overall composite enterprise resilience ranking. Enterprise resilience combines the vulnerability to operational disruption with the ability to recover from such disruption).

Some Vulnerable Countries

Next, we looked at economically prominent countries with both significant natural hazard exposure and high urbanization rates. These major global manufacturing hubs were among the vulnerable: Bangladesh (apparel and textiles), Thailand (electronics, automotive components and related equipment), Vietnam (electronics), China (raw materials, electronics and related equipment) and India (automotive and related equipment). The combination of high natural hazard exposure and a high urbanization rate heightens the potential for severe disruption across business operations and global supply chains with operations or partners in these regions.

Sweden (vehicles and machinery, including computers) is an example of a country showing a low natural hazard exposure and low urbanization rate. Singapore (refined petroleum, integrated circuits and computers) is an example of a country with high natural hazard exposure yet low urbanization rate. South Africa (iron, steel, motor vehicles, metals) is the opposite, with a low natural hazard exposure and high urbanization rate.

Urbanization rate and natural hazard exposure are just two of 12 drivers of overall resilience in our 2017 FM Global Resilience Index. Though designed with business decision makers in mind, anyone can visit the index and sort 130 countries and regions by any driver (such as natural hazard exposure or urbanization rate) or by a country's composite ranking indicating their relative overall enterprise resilience to disruptive events.

If, after looking at all of these rankings, you find yourself or your business is operating in a vulnerable region, it may be unlikely you can pack up move away. But you still have a chance to make your operations more resilient by fortifying it against those hazards to which it is exposed. And if you haven't committed yourself to a geography–and can settle somewhere that's less vulnerable–the Resilience Index is worth a look.

When it comes to the consequences of natural disasters, your fate is largely in your own hands.

The views expressed are those of the author(s) and are not necessarily those of Scientific American.

About the Author

Lou Gritzo is vice president of research at FM Global, one of the world's largest commercial property insurers.

"Positive Cities" Can Improve Earth as Well as People's Lives

By William McDonough

Cities are home to more than half the world's population, and they exert increasing stress on the earth. They produce up to 70 percent of global carbon dioxide emissions, use up vast quantities of water, degrade water quality and produce mountains of waste. As cities go, so goes the planet. And cities are growing–fast. By 2030, according to the latest United Nations estimates, five billion people will live in cities, nearly half of them conducting their lives in homes, schools, workplaces and parks that do not yet exist.

The challenges in making cities as sustainable as possible are enormous. But they are also inspiring because cities can play an outsized role in creating solutions for a more sustainable world. Cities are engines of innovation and entrepreneurial energy. As networks of city leaders are showing, municipalities are also powerful actors, pooling their strengths, setting environmental agendas and exercising global leadership. From megacities to towns, mayors and city councils, investors, economists and planners are responding to the urgent need to redesign the basic elements of fast-growing cities from the ground up. How they reimagine the urban landscape and how they design growth will profoundly influence the future of all life on the planet.

Many cities are taking important steps to reduce air and water pollution. They are being "less bad." Efficiency in itself, however, is insufficient to move us to a positive future. If cities can also be effective and do "more good," for example, by converting waste back into nutrients for food production, cities can propel us toward a future we want, not just reduce the impacts we do not want. More good, not just less bad.

A clear vision for reimagining the city and its relationship to its surrounding countryside can be found in replicating the operating system of the natural world. In essence, natural systems operate on the free energy of the sun, which interacts with the geochemistry of the earth to sustain productive, regenerative biological systems. Human systems, including cities, that operate by the same laws can approach the effectiveness of living systems.

These laws can be distilled into three key principles: equate waste with food; maximize use of solar income; and celebrate diversity.

Waste does not exist in nature, because each organism contributes to the health of the whole. A fruit tree's blossoms fall to the ground and decompose into food for other living things. Bacteria and fungi feed on the organic waste of both the tree and the animals that eat its fruit, depositing nutrients in the soil that the tree can take up and convert into growth. One organism's waste becomes food for another. Nutrients flow perpetually in regenerative, cradle-to-cradle cycles of birth, decay and rebirth. Waste equals food.

Our cities are currently designed for linear flow. Biological nutrients (such as food and wood) and technical nutrients (such as metals and plastics) enter at one end and are used, then discarded. After waste is sifted for valuable recyclables such as metals, paper and certain plastics, it flows out the other end, headed for landfills or incinerators. The process is "take, make, waste." But just as we have redesigned certain consumer products to be disassembled, recycled or reused, we can design cities in a similar, circular fashion: take, make, retake, remake, restore.

In the circular city, wastes become resources. Consider sewage. In a linear city, wastewater-treatment plants process food residues and human waste—including the valuable minerals they contain, such as phosphate—and release the effluent into rivers as pollution. Farmers then buy more phosphate from Morocco or other distant lands to make new fertilizer to grow more food in soils that have lost those minerals. In the circular city, sewage-treatment plants become fertilizer factories. The carbon, phosphate and nitrogen flowing out of the facility are seen as potential assets for soil, not as liabilities for the nearest river. Circular cities mine the wastewater for phosphate and turn it into fertilizer for parks and vegetable gardens on city rooftops and for farms and forests surrounding the city. This process eliminates the need to buy more phosphate from faraway sources and transport it to the U.S. It also avoids the energy and carbon emissions involved in mining and transport. Ostara Nutrient Recovery Technologies in Vancouver is one of several companies pioneering the collection of the phosphate mineral struvite for fertilizer from sewage sludge.

Eliminating the concept of waste extends to all systems, so circular cities design incoming materials for "next use" instead of "end of life." For example, new technologies allow cell-phone circuit boards to be profitably processed in a clean facility, where all the rare earth minerals and precious metals are recovered for reuse in new electronic products.

A second key principle of nature, and what we might call positive cities, is that everything is powered by the sun–and sometimes, as in the Iceland city of Reykjavík, by geothermal power. Trees and plants manufacture food from sunlight–an elegant, effective system that uses the earth's only perpetual source of energy income. Buildings can tap into solar income by directly converting the sun to energy and by passively collecting the solar radiation for heat and natural lighting. Winds–thermal flows fueled by sunlight–can also be harnessed. Together solar, wind and geothermal energy can generate enough power cost-effectively to meet the needs of entire cities and regions, even nations. Cities such as San Francisco are already making significant progress toward meeting their commitments to run 100 percent on renewable energy in the next 10 years.

The third key principle–diversity–is found in all healthy ecosystems. Each organism has a unique response to its surroundings that works in concert with other organisms to sustain the system. Each organism fits in its place, and in each system the most fitting thrive.

Urban designers aiming for what fits attend carefully to local ecologies. They assess the geology, hydrology, vegetation and climate. They incorporate natural and cultural history. By combining this rich "essay of clues," designers discover appropriate patterns for the development of the landscape. By doing so, they create possibilities for positive growth that supports life.

Ultimately what we want is a city designed to allow people to live and work in the same neighborhood. If residents can decompose cell phones at a clean factory that fits within a city's ecosystem, there is no need to relegate the factory to special zones of bad behavior at the outskirts of town. A positive city eliminates the need for zoning driven by concern about unsafe or unhealthy activities. Factories can be in the middle of clean residential neighborhoods, providing jobs so people can live within walking or biking distance. That opportunity, in turn, greatly reduces the need for commuting and transportation–a huge waste of resources and of people's time. And if fresh, healthy food is grown on rooftops across the city, as Method Products' South Side Soapbox factory in Chicago is doing, not only can local organic waste be a resource for food-growing systems, but people working at the rooftop farms can live locally as well.

Imagine everything we make as a gesture that supports life, inspires delight and finds harmony with nature. Buildings operate like trees: they sequester carbon, make oxygen, distill water, provide habitat for thousands of species, and convert solar income into all the thermal and electrical energy they need–and sell excess power to the neighbors. Buildings with on-site wetlands and botanical gardens recover nutrients from wastewater and clean what remains for on-site kitchens and bathrooms. Fresh air, flowering plants and daylight are everywhere. Buildings and communities function as life-support systems.

With this vision in mind, we can imagine food and materials produced in the surrounding countryside, created with tools and technology made in the city. The city returns waste as a raw material that replenishes the system, putting the "re" back into resources. Everything moves in regenerative cycles, from city to country, country to city, in natural and cultural networks that circulate biological and technical nutrition–the hardware and software of the 21st century. The metabolism of a living, positive city allows human settlements and the natural world to flourish together. If we are to make our cities truly sustaining and beneficial for all, we need to take this as a literal, strategic truth that informs all our designs.

It is compelling to lay out principles for a utopian future. But can existing cities actually put them into practice today? Some recent industrial facilities are demonstrating how.

The renovation and expansion of the Ford Rouge Center in Dearborn, Mich., transformed the massive, historic car- and truck-manufacturing complex into a model of industrial sustainability. The master plan incorporated a 10-acre green roof–the heart of a system of wetland gardens, porous paving, hedgerows and bioswales. The project turned a 100-year-old industrial brownfield site into a thriving storm-water ecosystem that captures, cleanses and slowly releases water to the adjacent Rouge River in ways that support watershed health. Native killdeer returned to nest a week after construction was completed.

Another model is Sustainability Base, NASA's center for science and computing at its Ames Research Center in Moffett Field, Calif. The facility has the potential to provide all its heating, cooling and energy needs–and even an energy surplus–from solar and geothermal sources and a fuel cell with advanced energy-management controls. Wastewater is treated on-site.

The award-winning Park 20|20 development in Hoofddorp, the Netherlands, is another model. It is a diverse set of buildings and open spaces now being completed across 28 acres. People can access the park easily by air, rail, bus and bicycle. Green zones, plazas, public gardens and canal boardwalks provide a connection to the larger community. Each building's size, structure and orientation are optimized to capture the sun's energy and light. Buildings across the park integrate energy, water and waste-management systems, acting as a single organism.

Can we expand these successes to entire cities? Inspiration comes from some intriguing places. One of them is Curitiba, Brazil.

Curitiba began transforming in the 1970s, under legendary architect and urban planner Jaime Lerner, who was mayor three different times from the 1970s to the 1990s. During his first term

Lerner realized that the poor town of several hundred thousand inhabitants needed better public transportation. Because a subway or heavy rail system would cost far too much, he instead asked Volvo to make 270 Swedish articulated buses, done within the city, which gave residents jobs. The city hired locals to build aboveground, street-side bus shelters, or *tubos*, from which people could travel anywhere for a flat fare. Instead of riders paying with a token as they boarded a bus, a slow process, Lerner had them prepay when entering the *tubo* platform, so when the bus arrived, they could get on quickly, reducing loading time and making the entire system efficient.

As a result of Curitiba's rapid population growth, garbage was piling up in narrow alleyways where trucks could not retrieve it. Lerner created a program to instruct children how to separate trash, then sent them home to teach their families. In exchange for the sorted garbage, people were paid in bus tokens or fresh food; Lerner paid them in a currency for mobility. Suddenly, everyone was using mass transit. Today up to 85 percent of Curitiba citizens use the bus, and 90 percent recycle. Curitiba recycles 70 percent of its refuse—one of the highest rates in the world.

This innovative thinking went on and on. Instead of building a big central library downtown, the city created a network of 50 small "Lighthouses of Knowledge" throughout the neighborhoods so that every child was within walking distance of a library. Local builders, of course, constructed the brightly colored buildings. The libraries work with municipal schools and offer thousands of books and free Internet for citizens ages three to 80. These and other steps transformed the city to a live-work design.

Now imagine if Manhattan had this same kind of vision—with local food growing on the rooftops of hundreds of schools and hospitals, providing nutrition as well as local jobs. Kids could use their optical sensors known as eyeballs to sort trash, pull out the plastics, bring them to recycling centers so they do not end up in the ocean and then get paid in toys. Clean factories would reformulate the plastics into monomers that can be used again. Everything is powered by the sun, and materials flow in continuous cycles of biological and technical nutrition.

What we are after, for all people, is something I would call the "good life"—a life that is safe, dignified and creative. Positive cities are the places where that can happen. If they are designed and run on this principle, everything gets better. We have to insist on the rights of humanity and nature to coexist, to bring together the city and its surrounding countryside.

Cities are designed, but they are also organisms. As French anthropologist Claude Lévi-Strauss pointed out years ago, cities are "something lived and something dreamed." As makers of living places, we cannot help projecting ourselves onto the landscape. But as we dream of our ideal cities, as we conjure the human weft on the geologic warp of the land, we can begin to see more clearly the true character of the place we inhabit, its spirit. Then, as we shape the nature of our cities, we will be making places that celebrate both human creativity and a rich, harmonious relationship with the living earth. We will be forging a new geography of hope.

Referenced

The Upcycle: Beyond Sustainability—Designing for Abundance. William McDonough and Michael Braungart. Farrar, Straus and Giroux, 2013.

U.N. World Cities Data Booklet 2018: digitallibrary.un.org/record/3799524?ln=en

NASA Sustainability Base: www.nasa.gov/ames/facilities/sustainabilitybase

About the Author

William McDonough is an architect who has pioneered design for sustainable development. He is founder of William McDonough + Partners and McDonough Innovation and co-founder of the Cradle to Cradle Products Innovation Institute.

Sustainable Cities Put Waste to Work

By Michael E. Webber

On December 20, 2015, a mountain of urban refuse collapsed in Shenzhen, China, killing at least 69 people and destroying dozens of buildings. The disaster brought to life the towers of waste depicted in the 2008 dystopian children's movie *WALL-E*, which portrayed the horrible yet real idea that our trash could pile up uncontrollably, squeezing us out of our habitat. A powerful way to transform an existing city into a sustainable one—a city that preserves the earth rather than ruining it—is to reduce all the waste streams and then use what remains as a resource. Waste from one process becomes raw material for another.

Many people continue to migrate to urban centers worldwide, which puts cities in a prime position to solve global resource problems. Mayors are taking more responsibility for designing solutions simply because they have to, especially in countries where national enthusiasm for tackling environmental issues has cooled off. International climate agreements forged in Paris in December 2015 also acknowledged a central role for cities. More than 1,000 mayors flocked to the French capital during the talks to share their pledges to reduce emissions. Changing building codes and investing in energy efficiency are just two starting points that many city leaders said they could initiate much more quickly than national governments.

It makes sense for cities to step up. Some of them—New York City, Mexico City, Beijing—house more people than entire countries do. And urban landscapes are where the challenges of managing our lives come crashing together in concentrated form. Cities can lead because they can quickly scale up solutions and because they are living laboratories for improving quality of life without using up the earth's resources, polluting its air and water, and harming human health in the process.

Cities are rife with wasted energy, wasted carbon dioxide, wasted food, wasted water, wasted space and wasted time. Reducing each waste stream and managing it as a resource—rather than a cost—can solve multiple problems simultaneously, creating a more sustainable future for billions of people.

Pollution as Solution

Lessons about waste abound in history. John Snow, a London doctor, deduced that terrible cholera outbreaks struck London in 1848 and 1854 because public water wells were contaminated by sewage. Building sewers was an obvious solution, but political leaders rejected Snow's findings because his ideas did not fit prevailing ideologies and because the actions were deemed too expensive. Similar rejection is offered for today's climate scientists, who tell us that our waste is killing us, though in a much slower and less direct fashion, and that fixing the problem will require significant investments in new infrastructure. Snow was later vindicated as a hero (perhaps the same fate awaits our present-day scientists) after new leaders created ambitious public works projects to cram 1,200 miles of sewers into a crowded city of three million people, ending the cholera problem. The work also created the lovely river embankments that still stand as a key piece of London's urban environs and along which many people stroll.

Today just flushing the waste away is not enough, however. After we reduce it, we should close the loop and use the remainder again. First, limit waste, then put it to work.

This new thinking begins by redefining our concept of pollution. Raj Bhattarai, a well-known engineer at the municipal water utility in Austin, Tex., taught me a new definition for pollution: resources out of place. Substances are harmful if they are in the wrong place: our bodies, the air, the water. But in the right place, they are useful. For example, instead of our sending solid waste to a landfill and paying the bill, it can be incinerated to generate electricity. And the sewage for a million-person community can be mined for millions of dollars of gold and other precious metals annually for use in local manufacturing.

This idea fits with the larger concept of the so-called circular economy—where society's different actions and processes feed into one another beneficially. Simply put, waste is what you have when you run out of imagination.

Less is More

One obvious place to start reducing waste is leaky water pipes. A staggering 10 to 40 percent of a city's water is typically lost in pipes. And because the municipality has cleaned that water and powered pumps to move it, the leaks throw away energy, too.

Energy consumption itself is incredibly wasteful. More than half the energy a city consumes is released as waste heat from smokestacks, tailpipes, and the backs of heaters, air conditioners and appliances. Making all that equipment more efficient reduces how much energy we need to produce, distribute and clean up.

Refuse is another waste stream to consolidate. The U.S. generates more than four pounds of solid waste per person every day. Despite efforts to compost, recycle or incinerate some of it, a little more than half is still dumped in landfills. Reducing packaging is one way to lessen this volume while also generating other benefits. Big retailers such as Walmart, for example, have found that reducing packaging results in fewer trucks needed for shipping and more shelf space to display goods.

Wasted food is its own heart-wrenching issue. Despite famine and food scarcity in many places globally, Americans throw away 25 to 50 percent of their edible food. Food requires vast amounts of energy, land and water to grow, produce, store, prepare, cook and dispose–so wasted food leaves a significant imprint. Initiatives that have popped up in the U.S., such as the I Value Food campaign, and in the U.K. are a start toward solving this vital issue.

Putting Waste to Work

Once cities reduce waste streams, they should use waste from one urban process as a resource for another. This arrangement is rare, but compelling projects are rising. Modern waste-to-energy systems, such as one in Zurich, burn trash cleanly, and some, including one in Palm Beach, Fla., recover more than 95 percent of the metals in the gritty ash that is left by the combustion. Rural villages, such as Jühnde in Germany, create enough biogas from cattle and pig manure to heat or power a large portion of their homes. My research group at the University of Texas at Austin has demonstrated that a cement plant in New Braunfels, Tex., can burn fuel pellets made of unrecyclable plastics rather than coal, avoiding carbon dioxide emissions and impacts from coal mining.

Even trash that is put in landfills can provide some value. Cities can collect methane that rises as the waste decomposes, which is an obvious improvement over flaring (burning off) the gas or simply letting the methane waft up into the atmosphere, where it traps much more heat than the equivalent amount of carbon dioxide. Power generators can convert the collected gas into electricity. Vancouver's landfills extract the methane and burn it to heat nearby greenhouses that grow tomatoes.

Even then, landfills are still leaky. That inspired Vancouver, which has pledged to become the greenest city on earth, to give residents separate bins for trash and organic matter (food scraps, yard clippings and tree trimmings). Officials expect citizens to use them properly and deploy city inspectors to check that waste haulers are dumping refuse that is separated correctly. The city produces methane from the organic waste while generating solids known as amendments that can make soil more fertile. These solutions solve multiple problems at once—saving money for energy that would otherwise have been purchased, reducing the need for expensive landfilling and avoiding unnecessary use and damage of land—while improving agriculture.

Austin does something similar with its wastewater sludge, passing it through anaerobic digesters to make biogas it sells or uses on-site for generating heat. It converts the remaining solids into a popular soil amendment known as Dillo Dirt (a reference to the armadillo, one of its local creatures). The city earns money by selling the Dillo

Dirt, offsetting some of the cost of treating wastewater. Although composting is a growing and popular trend among residents–and one worth pursuing–doing it poorly can actually lead to more methane emissions. For Austin, it makes more sense for residents to put food scraps down the drain and through a grinder so that the city's industrial-scale harvesters at the wastewater plant can do the work of the composter but with greater efficiency.

Waste heat is another big opportunity. Harvesting it is difficult because low temperatures are hard to convert into electricity. NASA developed thermoelectric generators to do this on its spacecraft, but the technology is expensive and inefficient. Nevertheless, advanced materials that can more effectively convert heat to electricity are coming. A place to start is the hot wastewater that goes down the drain when we wash our clothes, dishes or bodies. Sandvika, a suburb of Oslo, has massive heat exchangers along city waste pipes that extract heat to warm dozens of nearby buildings or defrost sidewalks and roadways. By turning on heat pumps in the summer, it can use some of the heat to cool those same buildings. Vancouver liked the idea so much that it repeated the concept, using wastewater to heat hundreds of buildings and the Olympic Village.

Taking that idea further is the Kalundborg Symbiosis in Denmark, a leading example of closed-loop thinking. The industrial park has seven companies plus municipal facilities–centered on electric, water, wastewater and solid-waste facilities–that are interconnected such that the waste from one is an input for another. Pipes, wires and ducts move steam, gas, electricity, water and wastes back and forth to improve overall efficiency and reduce total wastes, including CO2 emissions. For example, wastewater from the oil refinery flows to the power plant, where it is used to clean and stabilize fly ash from coal combustion. The refinery also sends waste steam to Novo Nordisk, which puts the heat to work for growing about half the world's supply of insulin with bacteria and yeast. The entire park looks like a living, industrial organism. And it has demonstrated economic growth with flatlined or reduced emissions.

Data-Driven Decisions

Can the Kalundborg Symbiosis model be replicated on a larger scale, for cities worldwide? Yes, but only if we make cities smart. An industrial park is flexible because it has only a few tenants and decision makers, but a city has many individuals and organizations making independent decisions about energy, water and waste every day. Integrating them requires a cultural shift toward cooperation, boosted by advances in smart technologies. "Smart cities" will depend on ubiquitous sensing and cheap computing, compounded by machine learning and artificial intelligence. This combination can identify inefficiencies and optimize operations, reducing wastes and costs while operating all kinds of equipment automatically.

Thankfully, making cities smart is an alluring objective for planners who want to accommodate higher densities of people without diminishing quality of life. For example, in India, where population and public health problems are severe, Prime Minister Narendra Modi has announced his intention to convert 100 small and medium-sized municipalities into smart cities as a possible solution.

The "smart" moniker itself is an accusation that most cities are dumb. That accusation sticks because municipalities rife with waste seem to be operating blind. The U.S. National Science Foundation has just launched a major research initiative called Smart & Connected Communities to help cities make better use of data. That name, by the way, indicates that intelligence is not enough–interconnections among systems and people matter, too.

Smart cities rely heavily on big data gathered from widespread sensor networks and advanced algorithms to quickly gain insights, draw conclusions and make decisions on those data. Connected networks then communicate those analyses to equipment all across the city. Smart meters for closely tracking electricity, natural gas and water use by time of day, household and industrial appliance are an obvious place to start. Real-time traffic sensors, air-quality monitors and leak detectors are also at hand. The Pecan Street consortium in Austin is collecting data from hundreds of homes to learn how access to such data streams might help consumers change their behaviors in ways that reduce consumption while saving costs. Cities such as Phoenix and military bases such as Fort Carson in Colorado have pledged to become self-sufficient users of energy and water and net-zero producers of waste. Achieving those ambitious goals will require a lot of interconnected data.

Better transportation may give urbanites their first glimpse of a smart city's benefits by cutting wasted time. Reducing the footprint of transportation means cleaning up the fuels, making the vehicles more efficient, reducing trip distances and duration, increasing vehicle occupancy and cutting back on the number of trips. If people live close to their work, they can walk or bike or use mass transit. Studies show that building protected bicycle lanes leads to dramatically increased ridership, and because bicycles require so little space, compared with cars, they can reduce congestion on the roads.

A driverless city will also free up wasted space and time associated with parking. With shared or autonomous cars in constant motion instead of private cars that are parked at home and work, the number of parking spaces can be restricted dramatically, opening up wasted space and easing congestion further. Researchers at the Center for Transportation Research at the University of Texas at Austin used sophisticated models to determine that shared, autonomous vehicles would lessen the number of cars needed in a city by an order of magnitude and would cut emissions, despite causing a slight increase in total miles traveled because the vehicles would stay in motion. Instead of wasting their time driving, commuters can rest, read e-mails, place phone calls or conduct other business. That work can create economic value–and trim a person's office hours so he or she can get home earlier for dinner.

Making our infrastructure smarter is certainly the key to solving basic problems such as leaky water pipes. Identifying leaks should be easy if meters are distributed throughout a water system to track flows and readily pinpoint the amount and location of those leaks. Researchers in Birmingham, England, developed a system with tiny pressure sensors that use a small amount of power to frequently check for and detect leaks in water networks, a big improvement over the old technique of waiting for someone to call and complain that water is shooting like a geyser out of the road. And someday we might send smart robots down the pipes to repair the problems.

High-performance sensors will also let us find and predict natural gas leaks before accidents happen. Gas leaks are not only bad for the environment and a waste of resources but dangerous, as we see in headline-grabbing explosions in urban areas with aging infrastructure.

It is hard to know where smart, waste-conscious cities may arise. I imagine a likely candidate will be a Midwestern town with a million people or more that needs to reinvent itself because its economy was gutted decades ago. Indianapolis comes to mind, in part because it needs to rebuild water, wastewater and sewer systems based on bad decisions a century ago. The city has been investing in its downtown and is on the rise. Pittsburgh is leveraging its existing assets–a vibrant urban core, city pride, forward-looking leadership from Mayor William Peduto, the strength of Carnegie Mellon University and other hotbeds of innovation–to go from being defined by its smokestacks to being defined by its brainpower. Indeed, Uber launched its autonomous-vehicle service there. Columbus, Ohio, which is the state capital and home to a major university, is another place to look for cutting-edge experiments in becoming smart. The U.S. Department of Transportation recently awarded Columbus a $40-million grant to reinvent its approach to mobility.

Getting from Here to There

Turning profligate cities into places that reduce waste and reuse what is left will not be easy. Integrated R&D investments from the federal government have to be combined with practical policies from all levels of government. Unfortunately, R&D funding is in recent decline, and in the U.S., it may drop further under the Trump administration.

Investment has to be socially savvy as well. Studies show that R&D for smart cities has focused more on technology than what the citizenry needs. Done the wrong way, the benefits of a smart city might accrue to those who already have Internet connectivity and access to advanced technologies, which would only widen the technology gap on top of other socioeconomic divides.

Municipalities also need to help residents become smarter citizens because each individual makes resource decisions every time he or she buys a product or flips a switch. Access to education and data will be paramount. Connecting those citizens also requires collaboration and neighborly interactions: parks, playgrounds, shared spaces, schools, and religious and community centers–all of which were central tenets of centuries-old designs for thriving cities. The more modern and smart our cities become, the more we might need these old-world elements to keep us together.

About the Author

Michael E. Webber is a professor of energy resources at the University of Texas at Austin and head of the Webber Energy Group there. He is also chief science and technology officer at ENGIE, a global energy and infrastructure firm in Paris that operates the world's largest independent electricity company as well as large natural gas networks.

U.S. Cities Lose Tree Cover Just When They Need It Most

By Richard Conniff

Scientific evidence that trees and green spaces are crucial to the well-being of people in urban areas has multiplied in recent decades. Conveniently, these findings have emerged just as Americans, already among the most urbanized people in the world, are increasingly choosing to live in cities. The problem—partly as a result of that choice—is that urban tree cover is now steadily declining across the U.S.

A study in the May issue of *Urban Forestry & Urban Greening* reports metropolitan areas are experiencing a net loss of about 36 million trees nationwide every year. That amounts to about 175,000 acres of tree cover, most of it in central city and suburban areas but also on the exurban fringes. This reduction, says lead author David Nowak of the U.S. Forest Service (USFS), translates into an annual loss of about $96 million in benefits—based, he says, on "only a few of the benefits that we know about." The economic calculation involves several such benefits that are relatively easy to express in dollar terms—the capacity of trees to remove air pollution, sequester carbon, conserve energy by shading buildings and reduce power plant emissions.

Nowak and a USFS colleague, co-author Eric Greenfield, found tree cover had declined in metropolitan areas across 45 states. The biggest losses on a percentage basis were in Rhode Island, Georgia, Alabama and Nebraska, together with the District of Columbia. Only three states—Mississippi, Montana and New Mexico—saw increased metropolitan tree cover, all by "nonsignificant" amounts. The research team used Google Earth imagery to examine 1,000 randomly chosen points in each state for a before-and-after comparison over a five-year period, generally ending in 2014 or 2015.

Nowak attributes the losses to factors including development to accommodate the expanding urban population, natural aging and death of trees, storms (Hurricane Katrina knocked out a third of the shade trees in New Orleans), insect damage (emerald ash borer has killed thousands of ash trees in Detroit), individual property owners converting forests to lawns or other uses, and fire.

The USFS study expands on a 2012 study by the same authors that looked at 20 individual cities, and found 17 of them had experienced significant tree loss. The question is whether the back-to-back declines in the two studies represent a temporary dip, Novak says–"or is it a line that keeps going down? If it keeps going down, I think we're going to be in trouble. Cities will warm up, we might have more pollution, people will be more unhealthy."

Novak is referring to robust and rapidly accumulating evidence that urban trees help reduce blood pressure and heart rate, decrease stress and elevate mental engagement and attentiveness as well as improve a sense of safety, comfort and overall happiness, along with other social and physiological benefits. A stunning 2011 study–since questioned by another study but replicated by several more–found having good tree cover in the immediate vicinity of a pregnant woman's home sharply reduced the likelihood she would have a low-birthweight baby, possibly by lowering stress levels. "There's almost no public health, crime or environmental quality metric that you can look at that isn't made better by the presence of trees," says Deborah Marton of the New York Restoration Project, a nonprofit organization focusing on New York City's disadvantaged neighborhoods. "Trees are infrastructure, and they are the cheapest way" cities can make a difference in the lives of residents, Marton says.

But that message has not gotten through to many municipalities. "Too many people think that living in closer contact with nature is nice, it's an amenity, it's good to have if you can afford it," says William Sullivan, head of the University of Illinois at Urbana–Champaign Landscape Architecture department, who studies the effect of tree cover on urban crime and was not involved in the new study. "They haven't got the message that it's a necessity. It's a critical component of a healthy human habitat."

Urban tree planting programs—even the heavily promoted "million tree" campaigns taking place in many U.S. cities—have not kept up with losses. Some of these initiatives have failed because of inadequate funding or fading interest. The American Bar Association, for instance, ended its nationwide million tree program in 2016 with only 52,693 planted. Others met their targets but the saplings are still too small to be picked up by Google Earth images; Nowak notes these recent efforts may pay off in a few years as the young trees mature. Still other organizers reached the million mark, "but took no interest in caring for the trees over time," Marton says. "It's slow. It's not sexy. If you plant a new tree, *that's* exciting. If you water it for five years... maybe it will grow a few inches." American culture, she says, is "all about the quick fix."

Other experts suggest cities are becoming more attuned to the value of trees, in part because of climate change. "Right at the moment, the urban heat island effect seems to be the most persuasive" for city governments, says Eric Sanderson, a Wildlife Conservation Society ecologist who focuses on urban habitats and did not take part in the USFS work. Densely developed areas typically experience higher temperatures. But shade from trees, together with evaporation of the water their leaves transpire, can help reduce peak summer temperatures in their vicinity by 2 to 9 degrees Fahrenheit (1 to 5 degrees Celsius). In Davis, Calif., researchers found shade trees reduced the surface temperature of asphalt by up to 36 degrees F (20 degrees C), and of the passenger compartments of parked cars by 47 degrees F (26 degrees C). For elderly people in homes without air-conditioning, the cooling effect of trees can make a life-or-death difference.

Some cities are also using trees to retain water and reduce runoff. Philadelphia, for instance, faced a U.S. Environmental

Protection Agency mandate to deal with the problem of storm water overwhelming sewage treatment plants. Officials considered building an underground tunnel as a temporary reservoir for the overflow. But after subtracting construction costs, a consultant estimated this approach would produce a net benefit over 40 years of just $122 million. Instead the city opted—with EPA approval—to plant trees and install other green infrastructure to soak up rainwater and keep it from getting into the sewer system in the first place. The same consultant projected this approach would yield a net $2.8-billion benefit in the form of improved property values, increased recreational opportunities and avoided heat-stress deaths.

Finally, pressure for environmental justice is also likely to drive the urban tree planting movement—especially because heat-stress deaths, low-birthweight babies and other effects associated with the absence of tree cover now tend to be concentrated in disadvantaged neighborhoods. Cities have plenty of room to improve. New York City has room for another 200,000 street trees, according to Jennifer Greenfeld, the city's assistant commissioner for forestry, horticulture and natural resources. And a study of urban California estimated that it has 236 million vacant tree sites.

Adding tree cover will, however, require a shift to long-term thinking—especially to plan ways to make room for nature while also accommodating new growth. "We're urbanizing like crazy," says Sullivan, the landscape architect. "And it takes a lot more than a few cities with million tree programs to replace the trees that get chewed up by office buildings and big box stores and parking lots." Living well in an increasingly urbanized world, he adds, will require "nature at every doorstep. It's not enough to have a phenomenal world-class park three miles from your home. It's not enough to have these incredible national parks five states away." A tree needs to grow, he says, outside every window and doorway.

About the Author

Richard Conniff is an award-winning science writer. His books include The Species Seekers: Heroes, Fools, and the Mad Pursuit of Life on Earth *(W. W. Norton, 2011).*

Cities Need to Prepare for Water Day Zero

By The Editors

Earlier this year ominous headlines blared that Cape Town, South Africa, was headed for Day Zero—the date when the city's taps would go dry because its reservoirs would become dangerously low on water. That day—originally expected in mid-April—has been postponed until at least 2019 as of this writing, thanks to water rationing and a welcome rainy season. But the conditions that led to this desperate situation will inevitably occur again, hitting cities all over the planet.

As the climate warms, extreme droughts and vanishing water supplies will likely become more common. But even without the added impact of climate change, normal rainfall variation plays an enormous role in year-to-year water availability. These ordinary patterns now have extraordinary effects because urban populations have had a tremendous growth spurt: by 2050 the United Nations projects that two thirds of the world's people will live in cities. Urban planners and engineers need to learn from past rainfall variability to improve their predictions and take future demand into account to build more resilient infrastructure.

How did Cape Town get into a Day Zero situation? The city gets its water from six reservoirs in Western Cape province, which usually fill up during the rainy season, from May through August. But since 2015 the region has been suffering from the worst drought in a century, and the water in those reservoirs dwindled perilously. Compounding the problem, Cape Town's population has grown substantially, increasing demand. The city actually did a pretty good job of keeping demand low by reducing leaks in the system, a major cause of water waste, and has even won awards for its conservation policies. But the government of South Africa was slow to declare a national disaster in the areas hit hardest by the drought, paving the way for the recent crisis.

Cape Town is not alone. Since 2014 southeastern Brazil has been suffering its worst water shortage in 80 years, resulting from decreased rainfall, climate change, poor water management, deforestation and other factors. And many cities in India do not have access to municipal water for more than a few hours a day, if at all. For example, the city of Shimla ran out of drinking water in May, prompting locals to beg tourists to stay away from the popular Himalayan summer retreat. The water infrastructure in many Indian cities is old and leaky, but city governments have not repaired it. Municipalities have, however, given free electricity to farmers for irrigation, depleting local groundwater stocks.

In the U.S., the situation is somewhat better, but many urban centers still face water problems. California's recent multiyear drought led to some of the state's driest years on record. Fortunately, about half of the state's urban water usage is for landscaping, so it was able to cut back on that fairly easily. But cities that use most of their water for more essential uses, such as drinking water, may not be so adaptable. In addition to the problems that drought, climate change and population growth bring, some cities face threats of contamination; crises such as the one in Flint, Mich., arose because the city changed the source of its water, causing lead to leach into it from pipes. If other cities are forced to change their water suppliers, they could face similar woes.

Fortunately, steps can be taken to avoid urban water crises. In general, a "portfolio approach" that relies on multiple water sources is probably most effective. Cape Town has already begun implementing a number of water-augmentation projects, including tapping groundwater and building water-recycling plants. Many other cities will need to repair existing water infrastructure to cut down on leakage.

Metropolitan leaders should be thinking about meeting long-term needs rather than just about daily requirements. Good organization and financial accountability are equally critical. And planning efforts should include diverse stakeholders from the community. One major challenge is providing services to informal areas, which develop haphazardly, without any government foresight. Such regions often lack basic resources—a well-planned water supply among them.

The global community has an opportunity right now to take action to prevent a series of Day Zero crises. If we don't act, many cities may soon face a time when there isn't a drop to drink.

Sustainability Solutions for Megacities

By Augustine Quek and Yen Wah Tong

Urban populations have grown unabated since the advent of cities. Some urban centers have such high populations (> 10 million) that the term, megacity, is used. The impact of megacities on the natural environment has reached unprecedented scales in human history. Growing resource and energy consumption, with corresponding waste generation, has exerted a toll on the regional and planetary environment.

Overexploitation of natural resources, air and water pollution and solid waste mismanagement are undermining sustainability in many cities. For example, the United Nations estimated that cities consume about 75 percent of globally generated power, which produce 70 percent of anthropogenic greenhouse gas emissions.

The problems are expected to exacerbate by 2050, when two-thirds of the world's population is projected to live in cities. Therefore, sustainable urban solutions are required.

Researchers of various disciplines in the Energy and Environmental Sustainability Solutions for Megacities (E2S2) program–a collaboration between Shanghai Jiao Tong University, in China, and the National University of Singapore–have conducted research and data collection in two cities of different size and complexity, the megacity of Shanghai and land-scarce Singapore.

A major problem in both cities involves the incineration of food waste, which comprises more than 22 percent of incinerable waste but is only 16 percent recycled in Singapore. E2S2 found solutions that coupled the problem of waste management with energy and material production.

These include a high-efficiency three-stage anaerobic digester (AD) that converts food waste to biogas, and a high-solids anaerobic digester that is water-efficient. We have developed gasification technology that can convert solid wastes and the digestates from the AD process into syngas and biochar. Pilot tests are now being run at eateries in Singapore to prove the feasibility of AD in an urban environment.

Other solutions include technologies for using low-quality waste heat in adsorptive cooling and dehumidification, wireless air-quality sensor network and monitoring of emerging contaminants in reservoirs in real-time, by a network of water drones. Studies were also done on toxicology and risks assessments for contaminants in air, water and land environments in the city.

These sustainable solutions are applicable for Singapore and other megacities in Asia and around the world, to provide a clean and healthy urban environment, while maintaining environmental sustainability.

The views expressed are those of the author(s) and are not necessarily those of Scientific American.

About the Authors

Augustine Quek is Program Manager for the Environmental Research Institute (NERI) at the National University of Singapore (NUS).

Yen Wah Tong is Associate Professor of Chemical and Biomolecular Engineering at the National University of Singapore (NUS).

What Is a Smart Grid, and How Might One Protect Our Energy Future?

By Michael Tabb, Jeffery DelViscio, Andrea Gawrylewski

What would you do if your electricity was suddenly cut? What if you didn't have power for days or weeks? What if no one in the city or the state you live in had any power either?

The possibility of a massive, extended blackout may sound far-fetched, but recent events show there are reasons for concern.

In February 2021, an electric grid failure in Texas caused more than 4.5 million homes and businesses to lose power. Months later, wide-scale droughts and wildfires threatened the western US with rolling blackouts.

In 2021 the American Society of Civil Engineers gave our energy sector a C– grade. It stated that the "majority of the nation's grid is aging, with some components over a century old—far past their 50-year life expectancy—and others, including 70% of transmission and distribution lines, are well into the second half of their lifespans."

But what is the power grid exactly? And how might making it "smart" help to protect us all from the shock of widespread power failure?

The grid is a chain of interconnected machines spanning North America. Power moves down that chain from generators (such as turbines and solar arrays) to very high-voltage transmission lines, to lower-voltage distribution networks, to your outlets, lights and electric-vehicle-charging station.

Grid operators have to constantly monitor supply and demand to make sure they're equal: if there's an imbalance, it can end up damaging power plants. To avoid that, utilities can respond in various ways, like so-called load shedding: automatically disconnecting some people's power. Or if supply and demand get really misaligned, generators may turn off to avoid damage, which can lead to a chain reaction of "cascading outages," or a blackout.

With new energy sources such as wind and solar power, maintaining a steady balance gets even more complicated. This is because power from these sources is inconsistent.

"Smart grids" sense small imbalances faster, making adjustments so problems don't get larger. They use sensors to constantly measure the status of different parts of the grid, and a series of devices that control the current flowing through different points. The sensors and controllers are run by computers that make automatic adjustments to maximize efficiency.

Before going too deeply into how our electrification system can evolve, it helps to know where it started.

The U.S.'s grid was born on September 4, 1882, at 3 P.M. Thomas Edison himself threw the switch.

He built the first central power plant in the country at 255–257 Pearl Street in New York City. The small grid it powered in lower Manhattan initially served about 80 customers, providing electricity to 400 lights.

Edison was devoted to direct current, or DC, where power flows in one direction. But his DC system had limitations because of its low voltage, which meant that the power station could only move current to lights up to a half-mile away.

Famous inventor and scientist Nikola Tesla thought Edison was wrong. He pushed for an electric grid powered by alternating current, or AC, where power regularly flows back and forth across a line.

AC grids are hard to set up because every generator has to be perfectly synchronized, producing electricity with equal frequency, voltage and phase.

The two would be locked for years in what was called "war of the currents."

Tesla won in the end, aided by the invention of electrical transformers, which transform AC to a higher voltage for more efficient transport over long distances and then reduce the voltage for safer household use. This was much harder to do with direct current.

Most urban regions in the U.S. had power by around 1920. For the vast majority of rural Americans, this did not begin until the

1930s. Because it was so expensive to build power plants and lines, state governments let power companies form monopolies over large areas in exchange for connecting everyone to the grid.

During World War I, when some cities overwhelmed their generation capacity with military production, regions started connecting their power systems to pool resources. That linking process continued until North America ended up with five power grids: eastern, western, Quebec, Alaska and Texas, which chose to keep its grid separate.

Starting in the 1970s, Congress began to open power lines to companies that could generate cheaper electricity. And many utilities started getting their power from more distant sources.

The delivery chain was never built to handle these loads, and no one had an incentive to upgrade shared power lines. This became a big problem by the early 2000s.

In 2003 overgrown tree branches knocked out some power lines in Ohio. The failure cascaded, and 50 million people lost power.

That emphasized the need to improve real-time monitoring across the grid—which would help check for the warning signs of an impending blackout so the system could automatically prevent failures or contain them to small areas.

In 2007 Congress called for developing a "smart grid." Since then, it has spent billions of dollars on new infrastructure with the ultimate goal of fully deploying a smart grid system.

In addition to more sensors and controllers, many smart grid models use a microgrid structure. This design can supply power to small communities while being interconnected to the larger network. When a line goes down, the microgrid around it can be isolated, so the blackout stays within that small area instead of cascading more widely.

Smart grids may also incorporate new long-distance networks that use direct current—reversing the century-long dominance of AC. As engineers have found new ways to build high-voltage power lines, direct current has actually become more efficient over longer distances than before. And renewable energy sources tend to generate direct current anyway.

More reliance on renewable energy also increases the need for big storage batteries that can quickly be turned on and off to balance supply and consumption even when it's not sunny or breezy enough to meet demand.

A smart grid, though, can only do so much. Engineers also need to make individual generators more resistant to disaster, especially because severe weather is happening more often, thanks to climate change. Individuals can also make their homes more energy-efficient.

Smart grids have some vulnerabilities. By connecting all of our electrical systems to computer networks, many of which are designed to be controlled remotely, security experts say we've placed ourselves at high risk for potentially disastrous cyberattacks.

Russian hackers have managed to break into American power companies at least once. And utilities are still trying to figure out how extensively their systems may be compromised.

But we shouldn't necessarily blame smart grids. Hackers can target generators even if they aren't directly connected to the Internet.

Once complete, a smart grid should be much more secure than our current infrastructure, given its focus on reliability and backup sources. Such grids promise a more reliable and economical electrical infrastructure, hardened against outside threats.

Our daily digital life has become even more power-hungry, given the ever increasing number of connected devices we use. Most of us probably don't think much about where that power comes from. Losing it suddenly—and unexpectedly—would come as quite a shock indeed.

GLOSSARY

acute When an issue appears suddenly and with great severity, rather than persisting over time.

adsorptive Capacity for liquid to adhere to the surface; different from absorptive.

disparate When two or more things are different enough that they cannot be compared.

favela Term for a slum or working-class area in a Brazilian city.

fiat Formal order or decree by an authority.

GDP Gross Domestic Product, a measure of the total value of a country's goods and services.

gentrification The changes in a neighborhood caused by wealthier inhabitants moving in.

globalization The process whereby the scope of economic activity, government, communications and/or culture takes on an international scale.

hierarchical Organized by rank, with the top rank having decision-making power.

informal economy Economic activity that is not recognized, regulated, or taxed by a government.

inimical Having a negative or antagonistic effect on something.

leverage Advantage someone holds during a negotiation process.

municipal Pertaining to cities.

palatial Large and grand, like a palace.

pseudoscience Beliefs that appear to be scientific, but are false and/or incompatible with the scientific method.

regenerative Self-sustaining and maintaining with minimal waste or inputs.

regression analysis An important method of statistical analysis that determines whether one measured variable is dependent on another.

sprawl The horizontal spread of built urban or suburban environments.

squatter Someone who occupies land or buildings that they do not own or rent.

sterile Lifeless, unimaginative or unfeeling.

traction The friction force that keeps one surface joined to another, like a wheel on a road.

undulating Smoothly curving.

urbanization The movement of populations from rural areas to cities over time.

walkable An urban planning term that assesses how easily someone can traverse a city on foot; incorporates density, safety, transit availability, and other measures.

FURTHER INFORMATION

"What Is New Urbanism?" Congress for New Urbanism. https://www.cnu.org/resources/what-new-urbanism.

Anderson, Maggie, and Lydia Chain. "In Lagos, Vulnerable Communities Are Buried by Urbanization." *Scientific American*, December 17, 2021, https://www.scientificamerican.com/article/in-lagos-vulnerable-communities-are-buried-by-urbanization/.

Biello, David. "The Start-Up Pains of a Smarter Electricity Grid." *Scientific American*, May 10, 2010, https://www.scientificamerican.com/article/start-up-pains-of-smart-grid/.

Brook, Daniel. "LEED Compliance Not Required for Designing Green Buildings." *Scientific American*, September 1, 2008, https://www.scientificamerican.com/article/leed-compliance-not-required/.

Huler, Scott. "Walk This Way." *Scientific American*, February 24, 2012, https://blogs.scientificamerican.com/plugged-in/walk-this-way/.

Kelley, Alexandra. "US Teams Up with Partner Nations to Release Smart City Cyber Guidance," Nextgov, April 20, 2023. https://www.nextgov.com/cybersecurity/2023/04/us-teams-partner-nations-release-smart-city-cyber-guidance/385412/.

Lamster, Mark. "Castles in the Air." *Scientific American*, September 1, 2011, https://www.scientificamerican.com/article/castles-in-the-air/.

Ly, David. "On the Horizon for Smart Cities: How AI and IoT Are Transforming Urban Living," *Forbes*, April 7, 2023. https://www.forbes.com/sites/forbestechcouncil/2023/04/07/on-the-horizon-for-smart-cities-how-ai-and-iot-are-transforming-urban-living/?sh=501cec7d7145.

CITATIONS

1.1 Myths of the City by Roger Doyle (September 1, 2005); 1.2 Lost Garden Cities: Pre-Columbian Life in the Amazon by Michael J. Heckenberger (October 1, 2009); 1.3 Gigalopolises: Urban Land Area May Triple by 2030 by David Biello (September 18, 2012); 1.4 Decoding Mexico's City of Gods by Erik Vance (July 1, 2014); 1.5 A Tale of Two City-States by Tali Trigg (March 31, 2015); 1.6 The Rise of *Homo verticalis* by Roland Bouffanais and Sun Sun Lim (January 16, 2019); 1.7 Cities Are Surprisingly Fragile by Michael Hunter, J. David Selby, and Kevin C. DeSouza (June 25, 2019); 2.1 Bigger Cities Do More with Less by Luís M. A. Bettencourt and Geoffrey B. West (September 1, 2011); 2.2 Grid Unlocked: How Street Networks Evolve as Cities Grow by Sarah Fecht (April 6, 2012); 2.3 Jane Jacobs: Is There Good Science behind Urban Planning? by Sarah Fecht (December 7, 2012); 2.4 How Traffic Jams Decentralize Cities by Sarah Fecht (November 20, 2013); 2.5 Simple Mathematical Law Predicts Movement in Cities around the World by Viviane Callier (October 1, 2021); 3.1 Why Building Green Can Keep People Out of Jail by Majora Carter (September 1, 2008); 3.2 Share the Wealth: New Urban Poverty Atlases Now Provide Data to Slum Dwellers by Larry Greenemeier (August 18, 2011); 3.3 Street Markets and Shantytowns Forge the World's Urban Future by Robert Neuwirth (September 1, 2011); 3.4 Race Colors New Residents' Views of Local Businesses by Erika Beras (December 2, 2015); 3.5 Photos Tagged as Art Linked to Rising Property Prices by Christopher Intagliata (May 2, 2016); 3.6 Who Benefits from Public Green Space? by Mallory Richards (September 19, 2020); 4.1 Revolutionary Rail: High-Speed Rail Plan Will Bring Fast Trains to the U.S. by Stuart F. Brown (May 1, 2010); 4.2 Auto Immune: Cities Convert Streets into Pedestrian, Cyclist and Mass Transit Thoroughfares by John Matson (June 15, 2010); 4.3 Can Suburbs Be Designed to Do Away with the Car? by David Biello (August 17, 2011); 4.4 Can a Sustainable City Rise in the Middle Eastern Desert? by David Biello (August 19, 2011); 4.5 The Inconvenient Truth about Smart Cities by Kendra L. Smith (November 17, 2017); 4.6 What's Holding Smart Cities Back? by Amit Nisenbaum, March 18, 2019; 5.1 The Fishy Business of Waste by Madhusree Mukerjee, April 1, 1995; 5.2 How Green Is My City by David Biello (September 1, 2011); 5.3 How Businesses Use Data to Outsmart Nature's Wrath by Lou Gritzo (June 13, 2017); 5.4 "Positive Cities" Can Improve Earth as Well as People's Lives by William McDonough (July 1, 2017); 5.5 Sustainable Cities Put Waste to Work by Michael E. Webber (July 1, 2017); 5.6 U.S. Cities Lose Tree Cover Just When They Need It Most by Richard Conniff (May 7, 2018); 5.7 Cities Need to Prepare for Water Day Zero by the Editors of *Scientific American* (August 1, 2018); 5.8 Sustainability Solutions for Megacities by Augustine Quek and Yen Wah Tong (July 10, 2018); 5.9 What Is a Smart Grid, and How Might One Protect Our Energy Future? by Michael Tabb, Jeffery DelViscio, and Andrea Gawrylewski (November 4, 2021).

Each author biography was accurate at the time the article was originally published.

INDEX